AF270090

EVERYONE BREAKS THESE LAWS

GERARDO CON DÍAZ

Everyone Breaks These Laws

HOW COPYRIGHTS MADE THE ONLINE WORLD

Yale UNIVERSITY PRESS   NEW HAVEN AND LONDON

Published with assistance from the Kingsley Trust Association Publication Fund established by the Scroll and Key Society of Yale College and with assistance from the foundation established in memory of Philip Hamilton McMillan of the Class of 1894, Yale College.

Yale University Press books may be purchased in quantity for educational, business, or promotional use. For information, please email sales.press@yale.edu (US office) or sales@yaleup.co.uk (UK office).

Set in Scala Pro type by IDS Infotech, Ltd.
Printed in the United States of America.

Library of Congress Control Number: 2024947874
ISBN 978-0-300-25126-5 (hardcover : alk. paper)

A catalogue record for this book is available from the British Library.

Authorized Representative in the EU: Easy Access System Europe, Mustamäe tee 50, 10621 Tallinn, Estonia, gpsr.requests@easproject.com.

10 9 8 7 6 5 4 3 2 1

Para mis padres,
Gerardo Con Sanchún y Dora Díaz Castro,
con mucho amor

CONTENTS

This book is about something most of us sense but rarely acknowledge, something that affects us every day in ways we may feel but can rarely describe. It is about how copyright laws, which give folks rights over their creative works, are woven into the fabric of the online world. Everything we do online ties back to copyright in some way, but few of us have an opportunity to understand how deep those connections are and why they exist.

Copyright law is slippery and flexible, and it is as flawed as any other human creation. Behind the legal maneuvering necessary for its enforcement are fascinating stories about hope, beauty, greed, and fear. Some of the things you take for granted about your online life hang on delicate threads of legal thought that could snap overnight. In these pages, I'll describe copyrights not just as legal tools to protect creative works but as testing grounds for core assumptions about our online privacy and self-expression. By the end of the book, I hope you'll see your online world as the latest version of a copyright system that powerful companies negotiate into being, a system that could have been different had one firm or one judge made different decisions.

The legal conditions that allow us to immerse ourselves in online media developed through the slow but steady erosion of people's ability to administer their own copyrights in rapidly changing technological environments. In the 1990s, creators could sue companies whose servers hosted or distributed their copyrighted works without permission, but now these companies are safeguarded by powerful laws, and legal responsibility falls mainly on users. Over the years, online platforms gained special protections that allow them to show endless streams of pictures without asking for permission, paving the way for the legal conditions that allowed today's generative artificial intelligence companies to flourish. I'll tell you how this happened, focusing on US law because that's my expertise and because so much online traffic travels through the United States. However, I hope you'll see that many of the things I'll show you cannot be contained by national boundaries.

The result of these changes is today's media-filled online world, enabled by a copyright system that makes life easier, less risky, and more profitable for large technology companies. But there's a price: less risk and more profit for the tech companies mean more risk and less compensation for the people who create the content in the first place, whether those creations consist of online porn, fine art, music, or investigative journalism. The emerging controversies over artificial intelligence have placed a spotlight on this situation, but the problem has been brewing for more than thirty years. Courts are now reconsidering some of the foundational ideas on which our online world depends. This book will help you to navigate what's coming—whatever that may be—by showing you just how tightly interwoven our online lives have become with the strange intricacies of copyright law.

EVERYONE BREAKS THESE LAWS

Introduction

IMAGINE THAT YOU WANT TO SEE pictures of cute dogs. You pick up your phone or laptop and type "cute dogs" into Google Images, and an endless stream of thumbnails floods your screen. Do this search on your own (or look at figure 1) and follow the breadcrumbs of the legal intricacies at play. Each thumbnail has some information that you probably ignore: a small logo for the website hosting the image and the first few words of the title of the page where you can find it. When you click on a thumbnail, you see a larger version of the picture, the full title of the website, a button to visit that site, and shaded-out text at the bottom of the frame: "Images may be subject to copyright. Learn more."

Our ability to find these dog pictures or any other creative work depends on the state of copyright law and its relations with online technologies. Copyright is the field of law that governs the ownership of creative works. It grants creators—such as photographers, writers, or singers—the right to exclude others from copying, selling, displaying, or otherwise distributing their works without permission. Most of the dog pictures that Google Images showed you are protected by copyrights. They belong to photographers or stock image companies, or to businesses that have had freelance photographers or employees take pictures of dogs. Whether or not these creators mind that we're looking at these pictures, Google is legally allowed to show them.

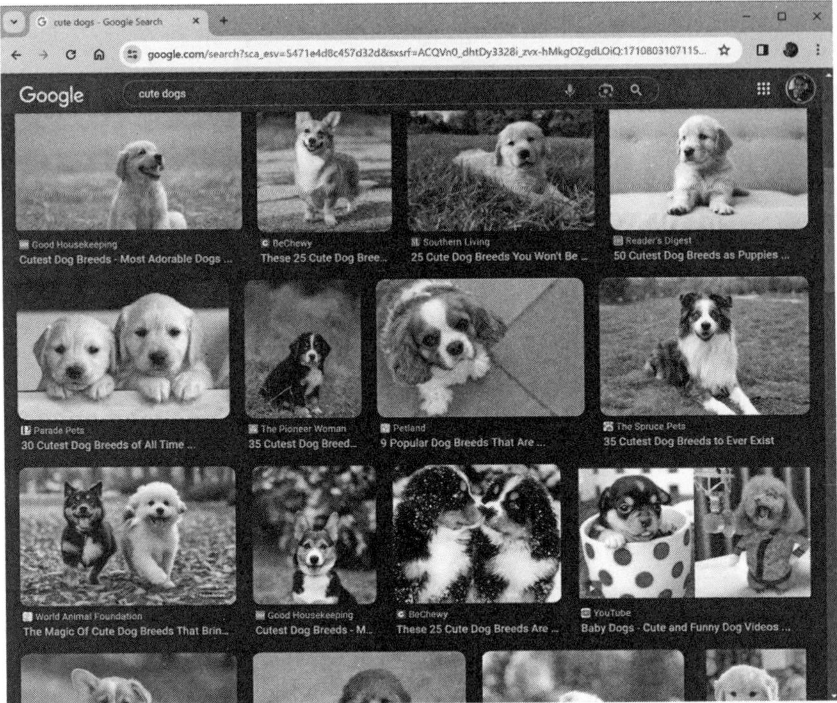

FIGURE 1. This is the results page I got after searching for "cute dogs" on Google Images. My use of this image, for scholarship and criticism, is protected by fair use. As we move along in the book, captions like this will show you different kinds of copyright arrangements that allowed me to show you the images.

This book is about the relations between copyrights and our online lives. I'll tell you stories about search engines, blogs, newsfeeds, digital music, artificial intelligence, and many other things that transformed what it means for us to create and access digital content online. Along the way you will get to know a very eclectic group of people, from Scientologists and porn librarians to brilliant scientists and rebellious hackers. By weaving these stories together (along with some of my own), I'll help you reflect on how copyrights have shaped the online world as *you* know it.

Copyrights are examples of something called intellectual property. IP grants people bundles of rights that let them exclude others from their creations. In addition to copyrights, IP includes patents (for inventions), trademarks (for brands and logos), trade secrets (for things a company needs to keep under wraps), and special protections for things such as plants and microchips. If you invent something, patent law grants you the right to be its only manufacturer; if you take a photograph, copyright law gives you the right to be the only one to duplicate it.

That doesn't mean, however, that everything human beings have ever made is protected by IP. Some of the dog pictures you found are in the public domain, meaning they are not protected by copyright. All IP protections are limited. Creative works made in the United States today are protected by copyrights for the life of the creator plus seventy years. And all copyrighted works are subject to an exception called "fair use" that allows unauthorized use for purposes such as parody, analysis, and criticism.

Fair use is very important: it's the reason I was able to show you the screengrab of cute dogs, even though I did not get Google's permission to show its website or the photographers' permission to show their work. It is a fundamental way of protecting free speech and ensures that folks can analyze and criticize creative works without fear of censorship. (Unfortunately, as you will see, it does not always work well.) Fair use has also morphed in the past decade or so to favor corporate interests over individual rights. It's a complicated situation that underlies both our ability to search for images online and the unauthorized use of artists' works by AI image and text generators.

Legal scholars and activists debate endlessly over the extent to which IP is essential to innovation and creativity. Some of the debate is quite polarized. At one extreme are folks who argue that all forms of IP improperly restrict knowledge and innovation. At the other extreme are those who insist that any weakening of IP rights is an affront to human inventiveness and creativity. More moderate positions tend to advocate a balance that is difficult if not impossible to reach: How do we ensure that creators and their employers receive the economic incentives they need to keep working? How do we do this in the face of technological change? What alternatives to IP do creators use, and in what settings are they sufficient?

I don't think we will find definitive answers to these questions or settle at one point on that spectrum. But those of us who think about IP regularly can agree on one thing: the history of IP and the histories of technological development and artistic expression are tightly linked. They influence each other partly by triggering changes in the broader commercial, technical, and cultural contexts that connect them.[1] Changes in how people create and disseminate their work can transform IP doctrines and practices. And new court rulings and bureaucratic procedures can enable firms and independent creators to commercialize products and compete with one another in entirely new ways. This is true of copyright but not unique to it.

Let's tie this all together by thinking about search engines again. Search engine result pages can reveal a lot about how human values filter into online platforms. For example, in her book *Algorithms of Oppression* (2018), Safiya Umoja Noble demonstrated that Google Images perpetuated anti-Black biases by showing primarily pictures of Black people when users entered terms such as "unprofessional hair" into the search bar. This was a surprise at the time, but it makes sense in hindsight: search engines rely on webpages' key words and tags to categorize images—and these terms, in turn, are chosen by human beings. This phenomenon is so important that much of the field of research known as data studies is dedicated to understanding the roles humans play in creating algorithmic bias.[2]

But when I think about search results pages, I first think about copyright law. Google did not get a license from each of the millions of dog photographers whose delightful work you can sample, and it doesn't need to do so because in 2007 it won a court battle against a pornography firm called Perfect 10. *Perfect 10 v. Google* established that Google was allowed to make thumbnails of other people's pictures and show them in its search results pages because this display counts as fair use. This ruling meant that whatever a photographer wanted to do with their images was irrelevant: if they are posted on a site that Google's search engine can reach, Google can legally include them in its search results.

Had *Perfect 10 v. Google* gone the other way—had the porn company persuaded the court that Google had no right to display its images without a license, and had higher courts agreed—then your search for cute dogs would have ended in frustration. Google Images would have had to ensure that all images on its results pages were either in the public domain or licensed by the company. Image search engines as we know them would simply not exist. But the implications of such a ruling would have been much bigger. Later in this book, I'll tell you how the ongoing controversies over AI art would also look quite different had the porn company won its case and why a monkey's selfie is important to that discussion.

WHY COPYRIGHT?

Almost every year at UC Davis, I teach an undergraduate course I call "Digital Law." We survey key computer-related developments in privacy and security, antitrust law, free speech, and IP. My students learn how law and information technologies have influenced one another in all these fields. And yet when I decided to write a book on this topic, I found myself writing mainly about copyrights. The reason is very simple: copyrighted works are everywhere. Think about how deeply immersed we are in other people's creative works. Perhaps you heard a song on your way to work or read a few articles on your phone while you waited in line for coffee. Maybe you scrolled through Instagram or TikTok posts while sitting at your desk or looked up a YouTube cooking video or recipe book at dinnertime.

Copyrighted works surround us because other people's creative output is embedded into our lives. But something deeper is going on: figuring out what it means to own something and how to govern different kinds of ownership is essential to our lives and economic systems. Michael A. Heller and James Salzman's book *Mine! How the Hidden Rules of Ownership Control Our Lives* (2021) explains this well for lay readers, but scholars have been thinking and writing about it for decades. Economic historians have shown how the legal systems that sustain IP regimes change in tandem with the push and pull of industrial and political forces.[3] Scholars in law and in science and technology studies have documented how ways of knowing the world have been inseparable from ways of governing it.[4] I also jumped into this bandwagon in my first book, *Software Rights*, which showed how the rise of software as a product and as a technology hinged on the legal, regulatory, industrial, and even personal politics of patent law.

One of the biggest lessons this scholarship has taught me is that conceptions of ownership can transform the social, cultural, legal, and economic environments in which they develop. For example, a law passed in 1998 called the Digital Millennium Copyright Act (DMCA) requires websites to have a mechanism for visitors to report "copyright infringement" (the unauthorized distribution or duplication of a copyrighted work). The name for these reports is a "DMCA Takedown." If an online service provider receives one of these, it must inform the person who posted the dubious content and give them a chance to respond. Service providers are not required to take down the content while they wait for the response, but if they don't, they open themselves up to legal liability.

DMCA takedowns are the bread and butter of US copyright enforcement in the online world. They are so common that Wendy Seltzer, a prominent legal scholar, founded a database called Lumen to document the staggering number of notices that sites receive from copyright holders (especially from big movie studies and record labels). You can visit it at www. lumendatabase.org. If you enter the name of a movie into Lumen's search bar, you'll find hundreds, if not thousands, of reports of takedown notices.

In recent years, however, DMCA takedowns have become something else: a tool by which people can censor the creators of content they find objectionable. There are penalties for filing a false notice, but it's easy to use a fake account that can't be linked to you. Angry readers with an urgent need to censor someone online thus have a surprisingly easy way to do so. A quick search on YouTube for "fake DMCA takedowns" will lead you to several videos explaining either how to file a takedown or how to respond to a fake takedown filed as an act of censorship.

This is a fascinating blurring between theory and practice. In the United States, IP protection is born from a clause in the Constitution that authorizes Congress to "promote the progress of science and useful arts, by securing for limited times to authors and inventors the exclusive right to their respective writings and discoveries."[5] This "exclusive right" is foundational to IP, and it provides creators with the financial incentives they need to keep creating, in the form of a guarantee that, for a limited time, they have the right to be the only ones monetizing their creations. And yet in the case of DMCA takedown notices we see how this right can become something else entirely: that a means to protect a financial incentive has turned into a tool to silence online speech.

Situations like these are what inspired me to write this book. I have three big goals. The first is to show you how the struggle to regulate the online circulation of creative works has eroded the cultural and legal boundaries that distinguish copyright enforcement from censorship. The second is to illustrate how this struggle has shifted legal responsibilities away from large corporations and toward individual users, challenging long-standing assumptions about what it means to make, share, own, and profit from a creative work. The third is to encourage you to reflect on how *you* have experienced these things, even if you have never noticed them.

When I say that copyrights *made* the online world, this is what I mean: efforts to carve out relations between copyright law and computer networks have played a central role in shaping the circulation of content across computer networks, what service providers can do with that content, and how users interact with it. I use the term "the online world" (singular) as an umbrella term. Online experiences unfold across several networks that do not necessarily connect with each other and that follow rules of access and behavior tied to the laws, cultures, and politics of the places where their servers and users are based.[6] People's lives and preferences take them to communities across these networks that many folks would never even want to access. Still, copyright law made *the* online world in a deeper sense—by creating rationales and rules for the transmission and display of content that transcended specific networks.

HOW TO READ THIS BOOK

Some questions about copyright never disappear from academic and policy circles. Do copyrights encourage folks to create more, or do they obstruct the free flow of human creative endeavors? Is copyright law flexible enough to handle both the needs of independent creators and the efforts of giant corporations? Would we be better off without any copyrights or with a stronger copyright regime? Are copyrights "good"?

We will encounter these questions repeatedly in the following chapters, but no single answer can capture the full range of opinions, practices, and interests involved. For this reason, instead of condemning or praising the system, I want to show you how copyright has become a flexible and evolving tool that people and institutions deploy and challenge according to their own social, political, cultural, and financial commitments. Over the past thirty years, stakeholders with competing (often incompatible) interests have slowly negotiated what it means to *be* online and what it takes for creative works to circle the globe in digital form.

This book is a collection of stories about people and communities running into the US copyright system in their everyday lives. Even the most ordinary things we do online hang on delicate legal balances and carry heavy historical baggage and dire implications. For that reason, I have divided the book into three parts. Their core stories, arranged in roughly chronological order, illustrate what it means to *break* copyright laws—not only in the formal legal sense but in the more colloquial sense of breaking something by straining or bending it too far.

If you appreciate academic labels, you can call this book a hybrid of science and technology studies with legal studies, grounded on the history of technology. But it is meant for anyone who is curious about why our online lives feel and work as they do. I chose these legal cases because they are central to the past or present of copyright law or else because they provide interesting or unusual illustrations. I focus primarily on the United States because it would be impossible to tell a global story without first considering the regimes in which giants such as Google and Facebook originated.

You can uncover copyright's grip on *your* online life by bringing yourself into the book: if you read a story about someone making an online post or trying out a new system, think about the times you have done something similar. Follow the links that start with web.archive.org to see what the web used to look like. (Please do so selectively with the awareness that sites may contain malware or other threats to your security.) Let the personal become analytical, and let stories connect with each other through experiences, speculations, and puzzles instead of just timelines, theories, or doctrines. You and I are part of these stories, too.

In early November 2004, when I was a college freshman, a journalist at the *Harvard Crimson* named Alexandra Bell asked if she could interview me. Alex lived downstairs from me at our dormitory in Harvard Yard. She was also a first-year student, and we enjoyed each other's company very much. She was my opposite in the best way possible: an outgoing British fashionista who was unbelievably socially savvy and read novels instead of math books.

Alex was writing an article about a new Massachusetts law that made it a felony to record movies at the theater using a video camera. She asked me if she could get a statement from me, and I agreed even though I knew nothing about movie piracy. Just days before my birthday, I gave her this embarrassing quote: " 'In Costa Rica it is a huge business with a lot of sourcing from states like Texas, Florida and California because they have more links to third-world countries,' said Gerardo Con Diaz '08, who bought pirated movies in his native Costa Rica, 'but I actually think there's almost no problem with bootlegging here in Massachusetts.' "[1]

This statement makes no sense. Why Texas, Florida, and California? What was I thinking, talking about links to "third-world countries"? And why in the world did I think that there was no bootlegging in Massachusetts? I had been in the United States for three months. Clearly, I just wanted to get my name in the *Crimson* so I could send the clipping to my parents.

I was proud of the quote for a few days and quickly forgot about it. But in the 2010s, when I was preparing to launch an academic career, I wished I could delete this article from the web. I investigated what it would take to do this and realized it would be a losing battle: even if I somehow persuaded the *Crimson* to take the page down, the site is preserved by the Internet Archive's Wayback Machine.[2] This is a service that periodically scours the web and makes permanent copies of websites. (It's also one of the resources I used to write this book.)

This was when I learned that in the battle to delete online content, copyright law has become an unexpectedly powerful weapon. If you find your own content on a website, the Digital Millennium Copyright Act of 1998 allows you to send the site a "takedown" notice: an official notice claiming ownership of this content. Site administrators must then contact whoever posted the material, and they often take down the content during the investigation to prevent themselves from being held liable. If I were determined to take Alex's article down and didn't mind committing fraud, I could file takedown notices at the *Crimson* and the Wayback Machine under the false claim that I wrote it.

Of course, I won't do any of that. Alex's article is a fun souvenir of my time in college, and I would rather not commit fraud. There's also the issue that getting in a legal brawl with one of the world's most powerful universities is risky business. But if I did have a legitimate reason to take down the article (perhaps if I had been its true author), I would be relieved to see that filing a takedown notice is very easy.

These notices are supposed to help copyright owners have agency in the online world. Some of the smaller sites have a page with instructions on where to email takedown requests, and bigger platforms have an intuitive interface where you can report content. At Google, the DMCA takedown procedure is part of the broader takedown system, which also includes takedown requests for such reasons as defamation and child pornography. You can try it out (and even file something) by visiting this site: https://support.google.com/legal/troubleshooter/1114905.

The process is so easy that you could even outsource DMCA takedown notices, though I would advise you to be cautious. The best way to do it is to hire an attorney. In practice, content creators turn to gig work platforms such as fiverr.com and upwork.com to find someone willing to do file notices for them. Recently I typed "DMCA" into fiverr.com and got 2,177 service offers from people around the world, starting at $10 for such tasks as filing a takedown notice. Many of these people offer their services to independent porn creators on such platforms as OnlyFans, which create newsfeeds of subscription-based content and give followers an opportunity to purchase access to more content.

A user called @ishtiaq_ahmad9, for instance, who lists himself as a "Dmca Attorney," offers a package deal for individuals who post their own pornography. He states: "If you are an adult content creator at onlyfans, Patreon, Chaturbate, Fansly or any other platform and people are leaking your premium content. If you want that this leaked content or stolen content should be removed permanently from websites and google search results. Then hire me to remove this all leaked content. I can remove leaked content under DMCA by sending takedown notices to google and other infringing websites."[3]

Based in Pakistan, @ishtiaq_ahmad9 claims to have successfully removed 155 instances of unauthorized content and to have several regular clients. He calls himself an attorney but doesn't list a law degree (an earlier version of the page listed a master's degree in mathematics from the University of South Punjab).[4] He seems to do very good work. He has thirty-three reviews averaging 4.9 out of 5 stars, many from independent porn creators.[5] According to a US client using the name maddycoles, "He was successful in removing leaked content from at least some websites directly, but all from search engines." Another US client, using the screenname maarya1, posted a screenshot checking off the URLs where @ishtiaq_ahmad9 had been successful.[6] These included fapomania.com/maarya-m and faponic.com/maarya, which he removed from both the hosting servers and their Google search entries.[7] Still pending were Google search removal requests for posts hosted at ilovnudes.com and camdenwhores.tv. Clients from Italy, Australia, South Korea, and the United Kingdom were also impressed by his work. The only dissatisfied customer (username cleosummers) gave him three stars and complained that the leaked content had not yet been taken down.[8]

Filing a takedown notice is so simple that the procedure is easy to abuse. Doing so involves at least one form of fraud, but anyone can do it with just a bit of technological savvy. Folks who want to protect themselves could file one using a VPN or rely on an international connection to create a fake digital persona. At this point, DMCA takedowns become something else: a weapon.

©©©

We're only starting to understand the extent to which the DMCA takedown notice is being abused. My research on this topic always brings me back to Wendy Seltzer's Lumen database, an online project that collects DMCA takedown notices and makes them available for study. I encourage you to visit Lumen's website and type in the name of your favorite television show or movie into the search bar. You'll be amazed at how many entries you find.

In 2022, for example, the legal scholar Shreya Tewari analyzed nearly thirty-four thousand seemingly fraudulent notices that Lumen received from June 2019 to January 2022.[9] All these notices used what is called the "back-dated article technique." Here is how it works: Suppose you find an online article with content you want to take down. You didn't write it and have no legal rights over it. You simply don't want it to exist. One thing you can do is create a new webpage, copy the original article onto it, and write on your page a publication date that precedes the original's publication date. Then you can file a DMCA takedown notice claiming that the original online article violates your copyright.

Tewari chose notices that referred to backdated articles published at a website that I urge you *not* to visit, today-news.press. This domain is owned by a private individual in Ulyanovsk, Russian Federation, and registered under a company that appears to be dedicated to owning URLs, called Registrar of Domain Names REG.RU LLC. The site is currently down, but you can check its status (or that of any other URL) safely by typing it into the search box at the following site: https://lookup.icann.org/en/lookup.

Sixty percent of the 33,988 notices Tewari found had been sent from the United States, and 28 percent from Ukraine or Russia. The rest came from France, the United Kingdom, Germany, and Switzerland. About thirty senders had sent these notices, targeting 550 online domain names. Their success rate was very low: just 0.8 percent led to an actual takedown.

We will never really know the full story behind these takedown notices, and I urge you read Tewari's blog post about them by following the note to this paragraph. She uncovered a fascinating censorship story: the targeted sites had also been banned by Russia's Ministry of Information Technology. Some of them contain serious criminal allegations against bureaucrats from the United States, Russia, and Kazakhstan. Others tie prominent people to the Russian mafia.[10]

Like many other legal means, DMCA takedown procedures are a tool. To understand their significance and power, we need to start by thinking about what the online world was like in the years surrounding the passage of the Digital Millennium Copyright Act. Before it became law, there was no unified procedure by which people could claim ownership of materials they found online. Instead, copyright owners targeted online service providers such as website hosting companies directly, threatening to shut them down unless they met the unreasonably difficult requirement of deleting all allegedly infringing material.

The legal and technological conditions of our online world have changed since the 1990s, but many ideas and practices from that period are still with us. You can see them whenever folks use copyright claims to censor others, when individuals and small businesses struggle to figure out the infringement charges against them, and every time you go online in search of casual entertainment.

CHAPTER ONE

Scientology Online

AT 7:30 A.M. ON FEBRUARY 13, 1995, a former minister of the Church of Scientology named Dennis Erlich called the 911 operator in Glendale, California, because a man in a suit would not stop ringing his doorbell. The operator told him that police had a search warrant to enter the home, so Erlich opened the door.[1] Two off-duty cops from the Inglewood Police Department came in, along with a reporter for the *Los Angeles Times*, an attorney for the church, someone Erlich described as a "scieno computer expert," and a high-ranking Scientologist named Warren McShane.[2]

I found a video of this event on YouTube and archived it so that you can watch it via the link in the notes.[3] The computer expert made a beeline for Erlich's personal computer, and everyone else started rummaging through his books and magazines. The visitors confiscated more than three hundred floppy disks, twenty-nine books, and one hundred megabytes of hard drive backups, loading hundreds of files into their own drives and wiping out the copies stored in Erlich's computer. The attorney gave Erlich a partial inventory of seized materials, and the team left as quickly as they had arrived.[4]

Erlich was shaken but not entirely surprised. Even though he was not expecting a raid, he knew it was only a matter of time before the church went after him. Since leaving the ministry, he had become one of Scientology's most vocal critics. His online posts revealed some of the church's most tightly held secrets, including direct quotations from confidential materials and sometimes complete documents. The Church of Scientology was no stranger to censorship. In the past, it had launched defamation lawsuits against its critics.[5] But in Erlich's case, the rise of home Internet pushed church leaders to deploy a new censorship weapon: copyright infringement lawsuits.

FAITH AND SECRECY

The Church of Scientology is often in the news, and rumors abound regarding its teachings and the influence it has over members and enemies. Celebrity Scientologists including Tom Cruise have raised the church's profile even higher, and I remember running into members frequently when I was in college. They offered free stress tests and advertised a book titled *Dianetics,* whose cover featured a picture of an exploding volcano.

As founded in 1953 by the former science fiction writer L. Ron Hubbard, Scientology has its own story about the origin of life on Earth. Ninety-five million years ago, at an intergalactic coalition of seventy-six planets called the Galactic Confederation, an evil ruler named Xenu faced an overpopulation problem. His solution, according to church writings, was to freeze immortal spirits called Thetans and send them to Earth. The Thetans' physical manifestations on Earth, humans, carry false memories designed to keep them in servitude.[6]

The church promises to help people identify and shed their immortal spirits' false memories to become "Operating Thetans." These are beings who, according to a study written around the time of Erlich's raid, "can act independently of their physical bodies and cause physical events to occur through sheer force of will."[7] To achieve this state, Scientologists use a device that resembles a lie detector to "audit" themselves, and they study Hubbard's writings.

Whether Scientology is a legitimate religion is an ongoing debate that I will not linger on here. The US Supreme Court ruled in the case of *United States v. Ballard,* in 1994, that the freedom of religion encoded into the Constitution guarantees that a religion's content is irrelevant to its recognition as such under the law.[8] It doesn't matter that Scientology has no deity or that its cosmology is a space saga published less than a century ago: a court recognized it as a religion in 1969 (though the Internal Revenue Service didn't grant it tax-exempt status until 1993).

Scientology has sacred texts. About seven hundred pages of Hubbard's confidential works cover eight levels of Operating Thetan that tell the full story of this intergalactic spiritual saga. The materials also provide detailed instructions on how church members can shed their false memories and unleash the full potential of their willpower.[9] The church treats these texts as technologies: they are formal inscriptions of Scientology's beliefs as well as objects that enable readers to accomplish a goal. The church sometimes calls Hubbard's texts its "Advanced Technology" materials. Treating (and managing) these texts as technologies is so important that in 1982 the church founded Religious Technology Center (RTC) to control all the texts and symbols of Scientology, including Hubbard's texts, as intellectual property.

Religious Technology Center's mission is "to espouse, present, propagate, practice, ensure, and maintain the purity and integrity of the religion of Scientology." Its articles of incorporation state that it is meant to "act as the protector of . . . Scientology" by enforcing the church's intellectual property rights, including trademarks and copyrights for major brands such as Dianetics. The church describes these texts as a "substantial body of confidential advanced religious technology which is a part of a body of truths and methods of application."[10]

By the late 1990s, the church was protecting Hubbard's writings through both copyrights and trade secrets.[11] This is unusual: intellectual property protection as a trade secret is normally granted to secret knowledge that engenders a competitive advantage in the marketplace. The recipe for KFC chicken, the algorithm behind the Facebook feed, and (more controversially) a firm's diversity-related demographic data are all protected trade secrets.[12] Upholding a trade secret in court requires its owner to demonstrate that the secret gives them financial gain and competitive advantage.[13]

This dual IP protection became an important weapon in the church's long-standing efforts to defend its reputation. As two legal scholars noted in 1997, defamation suits are difficult to win. They require showing that someone has acted out of "actual malice"—publishing information while knowing it was false or with "reckless disregard for the truth."[14] In addition, the Supreme Court had ruled in 1964, in the case of *New York Times v. Sullivan,* that "debate on public issues should be uninhibited, robust, and wide-open."[15]

The strong proofs demanded of plaintiffs in libel and defamation suits meant that the Church of Scientology had difficulty silencing its critics. In the early 1990s, after *Time* magazine published a cover story titled "The Thriving Cult of Greed and Power," the church sued parent company Time

Warner for defamation and a former member quoted in the story, Stephen Fishman, for libel. These suits remained in the courts for five years. The church was ultimately unsuccessful on all fronts, but Fishman endured several years of legal headaches.

By focusing on IP, especially on copyrights, the Church of Scientology no longer had to prove malice or falsehood and could focus on showing that its copyrighted works had been distributed without permission. I should note that Scientologists did not invent this legal maneuver. In his fascinating book *Copyrighting God,* legal scholar Andrew Ventimiglia shows how religious organizations have mobilized copyright law for more than 150 years and how this strategy has allowed them to create and police spiritual communities in ways that affirm, and sometimes reconfigure, their own ethical and theological commitments.[16]

But enforcing copyright law online would bring new headaches to Scientologists and critics alike, leading to such questions as: How do you stop the circulation of something that anyone can download and then re-upload? Who is to blame when copyrighted materials are found on a server? And the surprisingly difficult one: What *is* the Internet?

DIAL-UP INFRINGEMENT

The online world that Erlich was navigating was not the World Wide Web we know today. He was connecting to a bulletin board system (BBS) called support.com. These systems were dial-up online services that allowed users to connect to a limited online network where they could exchange files, play games, post and read messages, and chat with one another.

Bulletin board systems allowed online media to become social, well before the rise of today's social media. The media historian Kevin Driscoll has described this succinctly in his book *The Modem World:* "It was on these grassroots networks that computer enthusiasts started to use their machines for popular communication and community building."[17]

One of the communities Erlich frequented through support.com was an online message board called alt.religion.scientology (a.r.s). A critic of the church named Scott Goehring founded the group because he wanted "a place to disseminate the truth about this half-assed religion."[18] The group attracted supporters and critics of Scientology, and while disagreements were fierce, the group became a useful resource for folks who wanted to hear all sides of the debates.

In the mid-1990s, journalist Wendy M. Grossman published an award-winning article about this message board, titled "alt.scientology.war," and then a book, *Net.Wars,* in which she uncovered a series of seemingly coordinated attacks that the Church of Scientology launched on online communities.[19] But I'm not interested in retelling stories about the church. Instead, I want to uncover the legal twists and turns of Erlich's saga using a gold mine of court documents that an organization called the Electronic Frontier Foundation (which I'll discuss in a moment) archived online.[20]

Late in 1994, several posts criticizing Scientology were mysteriously deleted and replaced with a notice that read "Canceled because of copyright infringement."[21] System operators sometimes deleted users' posts, but the alt.religion.scientology case was suspicious because many of the deleted messages didn't contain any church materials. One deleted message, for instance, was a publicly available legal document from Spain, detailing criminal proceedings against the church's president and some of that country's most prominent Scientologists.[22]

alt.religion.scientology was not the only group having unusual encounters. By year's end, the managers for several other groups discussing scientology had received requests to take material down. Rumors (and documents) started to circulate among BBS users. The cancellation requests appeared to come from church members and from Scientology's security branch (the Office of Special Affairs International).

Erlich was determined to keep posting. After being excommunicated in 1982 and later being labeled a Suppressive Person (an enemy who impedes the church's progress), he had dedicated his life to debunking the church. He found it immensely frustrating that folks in groups such as a.r.s "were discussing this stuff like it was some sort of tea party."[23] As he wrote in one post, "They excommunicated me and now I'm making use of that same material by preaching or writing or publicizing my religious obligation onto the Internet."[24]

In 1994 and 1995, Erlich published his online critiques using the screen name "inFormer." He posted the texts of print articles, internal documents that the church considered private, and even some of Hubbard's sacred works. The internal documents disclosed obscure details about the church's organization and policies, its confidential beliefs, and the training that high-ranking members received as part of their initiation.

We will never know if church leaders were following a.r.s, but their lawyers certainly were. Leading the charge was Helena K. Kobrin. She worked closely with Thomas M. Small, RTC's lawyer and a member of the Glendale raiding party. Kobrin paired the complaint against Erlich with a direct message campaign aimed at shutting down alt.religion.scientology. Earlier that year, she had contacted servers hosting the newsgroup to say that the group's name was a trademark violation, that its earliest posts were forged, and that its main purpose was copyright and trade secret infringement.[25]

In addition to alleging that Erlich was posting copyrighted materials and trade secrets, Kobrin's complaint listed two more defendants: a man named Thomas Klemesrud and an Internet service provider (ISP) called Netcom Online Communication Services.[26] By including them, Kobrin was targeting the entire online infrastructure that allowed Erlich to participate in alt.religion.scientology. The goal was not just to stop Erlich but to create a legal precedent that would let the church intervene at key nodes in the flow of information across a network.

This is how Erlich's network access worked: Using his telephone and modem, he would connect to support.com, a BBS that Klemesrud operated. support.com hosted a.r.s, but Klemesrud did not have direct access to the Internet. He obtained it from Netcom, one of the largest ISPs at the time. In practice, this meant that Erlich would write a message, connect to the BBS, and upload it to Klemesrud's computers. Klemesrud would periodically connect to the Internet, upload all the messages that users had posted, and download everything that others had uploaded since the last connection. The system connected to USENET, a worldwide discussion network in which most groups, including a.r.s, did not have human moderators.[27]

Religious Technology Center did not raid Klemesrud's or Netcom's offices, but Kobrin was in touch with both for several months before the church filed its suits. This is how Klemesrud recalled the events: In December 1994, Kobrin called Klemesrud to demand that Klemesrud cut Erlich off from the BBS. Klemesrud told her not to call him again because he considered it harassment, and he hung up on her. In a follow-up email, Kobrin asserted that any claims Erlich made about fair use were false, accused support.com of divulging church trade secrets, and threatened legal action against both Klemesrud and Netcom if they failed to shut down alt.religion.scientology.[28] Klemesrud, a frequent contributor to a.r.s, responded that a BBS is "a distributor of information . . . much like a magazine rack, or bookstore." He added, "Here is what I can do for you. Anytime you see a posting on the support.com BBS, please mail me the original copyrighted work to compare. If I agree, and I will be fair, I will delete the material from the BBS. Public postings average around 3 days life span on the BBS."[29]

According to Klemesrud, Kobrin was not satisfied. The posts, she replied, included confidential church documents, so agreeing to this procedure could mean that she had to mail secret texts to a community of online critics. But Klemesrud insisted that this was the only way he could comply with takedown requests without having to cut users' access on demand, which he said amounted to "effectively curtailing their right to free speech."[30]

This response pushed Kobrin and her colleagues to cast a wide net in their legal complaints. Following Kobrin's lead, RTC sued Erlich for direct copyright infringement, namely reproducing, distributing, and displaying copyrighted works. It also accused Klemesrud and Netcom of direct infringement as well as contributory infringement—enabling others to infringe the church's copyrights.[31]

All these complaints reveal the difficult legal problems that home computer networks were starting to create. Should Klemesrud or Netcom be required to monitor every bit of information that went through their servers to ensure that no one's copyright was violated? And perhaps more important: If a user violated someone else's copyrights, should system operators be held legally responsible?

A TECHNOLOGICAL PIPE DREAM

On the unusually warm morning of February 21, 1995, Dennis Erlich arrived at the federal courthouse in San Jose, California, for a preliminary hearing. This is a gathering where a judge decides whether a full trial is needed. Sometimes judges issue temporary orders to the parties involved.

Erlich did not have an attorney. He was going to represent himself, but he was not alone. Netcom had brought two attorneys, and Klemesrud had brought one. All three were Bay Area lawyers with expertise in intellectual property law. They wouldn't defend Erlich, but they would at least be able to follow the copyright arguments that Kobrin and her lawyers had crafted.

The presiding judge was Ronald M. Whyte.[32] Appointed to the bench by George H. W. Bush, Whyte would soon become one of Silicon Valley's most important judges in matters regarding patent law. He had already sent

Erlich, Klemesrud, and Netcom temporary restraining orders that barred them from duplicating Hubbard's works, discussing any confidential materials, and transporting any RTC materials outside the court's jurisdiction. Now he needed to evaluate whether the court would issue preliminary injunctions against all three defendants—that is, orders prohibiting certain behaviors until the court decided the case. These orders would bar Erlich, Klemesrud, and Netcom from storing Hubbard's works in any hardware, databases, network facilities, or archives; displaying any works for which RTC held copyrights or causing them to be displayed; and destroying or concealing any of Hubbard's works, digitized or otherwise.[33]

Whyte had not yet read all the materials the court had received. One of the few documents he had read carefully was a statement by Rick Francis, Netcom's vice president for software engineering, in which Francis claimed that Netcom was merely an "Internet connectivity provider." Although it provided its clients full Internet access, he said, it was "not involved in any way in determining the content of the information available" online. This distinguished Netcom from other ISPs, such as CompuServe or America Online, which offered limited subscription-based plans that allowed users to access preset collections of pages and services.[34]

Francis's point was that an Internet service provider could not possibly monitor every file moving through its network. He estimated that 150 million keystrokes traveled through Netcom and into the Internet each day—150 megabytes of data. Handling this amount of data was especially difficult because it came from a vast network of clients. Klemesrud's BBS alone could access nine hundred newsgroups, and about two thousand other Netcom clients had this level of access.

Bulletin board systems allowed individual users to access the Internet without having to deal directly with Netcom. A BBS user would connect to a computer like Klemesrud's, which would then post the user's messages automatically to Internet newsgroups and forums. This made it unfeasible for Netcom to establish editorial control over individual BBS users: the only way for Netcom to shut down Erlich's access to the Internet would have been to terminate Klemesrud's access, disconnecting all of his five hundred subscribers.[35]

Francis's statement was enough to convince Whyte, early on, that "it's probably a practical impossibility" for Netcom and Klemesrud to "do any kind of censoring or checking" of what their users published or downloaded. This annoyed RTC's lawyers, who insisted that it should not be difficult to ban users such as Erlich without needing to disconnect anyone else. If messages contained information about their point of origin—that is, about who posted them—wasn't it possible, at least, to flag those that originated from Erlich's account?[36] They told Whyte: "We had a conversation with one of our programmers just the other day who was able in a fairly short period of time to write a program that allowed us to know when Mr. Erlich and other people were actually logging onto the internet. And we believe that it would be possible in very short order to create software that would allow the necessary monitoring to take place, and we're willing to work with defendants on that."[37]

Kobrin and her colleagues insisted that a technical solution was both necessary and possible. They emphasized that "computer science today would have the ability to flag those messages" and suggested that a widespread surveillance system could be implemented without much technical hassle.[38] But this was a dead end: the feasibility of such a system was not enough to convince Whyte that Netcom and Klemesrud could be compelled to do anything at all.

Of course, the technical fix Kobrin proposed was, and continues to be, a pipe dream. In his book *Wired Shut,* Tarleton Gillespie shows how during the 1990s and early 2000s, the history of digital rights management technologies was plagued by serious disagreements, even within individual industries, over the project's technical and cultural aspects.[39] Nowadays we have much more effective ways of managing digital rights, but the history is still unfolding. Platforms such as YouTube have automatic filters designed to flag copyrighted material, but they are not 100 percent effective. Even filters designed for primarily text-based materials are very easy to fool, as has been happening with websites designed to detect plagiarism.

In any case, Whyte was not inclined to issue a restraining order based on the promise of a future technology. Instead, he focused on the extent to which Erlich had violated the church's intellectual property rights. Religious Technology Center presented the case against Erlich as a straightforward finding of copyright infringement, but Whyte did not seem to think that Erlich was entirely at fault. Certainly, posting full documents without permission or a fair use defense would amount to copyright infringement, and disclosing the contents of secret materials could be a trade secret violation. But Whyte took issue with the church's attempt to treat all copyrighted work as secret.

At the start of the hearing, Whyte made it clear that he was not willing to support the use of copyright law as a means to censor Erlich. He stated: "What I'm concerned about is that he has a right, it seems to me, to do satires or articles or criticisms or anything he wants with respect to publicly available information pertaining to Scientology."[40]

This decision to protect Erlich's legal right to parody and criticize Scientology led to a troubling revelation: the church had made no distinction between publicly available materials protected by copyright and confidential texts that qualified as trade secrets. Instead, it had brought a claim against Erlich for copyright infringement even when he was doing things that would normally count as fair use. Religious Technology Center lawyer Thomas Small admitted that the church had not made those distinctions and even tried to argue that "paraphrasing of copyrighted works, close paraphrasing, is copyright infringement as well."[41]

Whyte asked Erlich to say more about what he had done, but this was hard for Erlich to do. He told the court that RTC's raid of his computers had been extremely thorough, and he no longer had direct access to his works. Perhaps other BBS users had saved copies of the posts, but as far as the court was concerned, RTC was the only party that could comment on the amount of copyrighted materials Erlich had used and how he had used them.

Erlich thought this situation showed how aggressively the church had targeted him. He told the court that RTC had taken his own copyrighted works from him, leaving no traces behind for him to mount a legal defense. "I have no yardstick to determine whether or not they're just trying to shut me up," he said, "or are they actually—do they have a case against me?"[42]

Erlich's lack of legal representation did not stop him from arguing that at least some of his actions were protected by fair use. He acknowledged having posted copyrighted materials but emphasized that he had done so as "a satirist, a writer, a minister." Some of his posts included "trade secret or whatever you want to call it," but he had never removed physical documents from the church without authorization (though some texts were sent to him anonymously). He also explained that he done this not for money but "to attempt to reform the organization that I thought had gone corrupt."[43]

This was not a legal argument for fair use, but Whyte got the point. The judge asserted Erlich's right to keep making online posts about the church but chastised him for posting extended excerpts or full texts with minimal commentary. He also extended a temporary restraining order to "prohibit any publication of confidential matter or any publication of copyrighted matter" that didn't qualify as "fair criticism or comment or fair use."[44]

In practice, this meant that until a full trial could take place, Erlich was free to keep posting and even quoting church materials—as long as he did not disseminate confidential information or publish, word for word, large portions of protected work. Erlich's battle was not over yet, but he was reassured that copyright law did not grant the church the right to censor anything he had to say about it.

METAPHORS AND THE LAW

Metaphors matter in law because they are how judges make comparisons between things they already know and new things awaiting their judgment.[45] All trials involve two sides pitching competing narratives against each other in an effort to negotiate what the "truth" is and tease out its legal implications. In common law systems such as the American one, metaphors allow judges to draw parallels between the situation at hand and the judgments their predecessors have handed down.

Time and again when researching the history of software patents, I have found that one of the major problems the courts faced in the 1990s was determining what software *is*—its nature as invention, technology, property, and commercial good.[46] No single definition satisfied all stakeholders, and many definitions contradicted one another, each one pushing courts in different legal and industrial directions. Software could be a machine, a text, an algorithm, or all or none of the above, and its nature could change from one moment to the next depending on the financial, intellectual, and cultural commitments of the person describing it.

A similar thing happened to the Internet, and not just in courts of law. For years, researchers have documented how metaphors enable communities and institutions to carve out a place for the Internet in their everyday lives.[47] But in the mid-1990s, especially at the federal level, a single metaphor dominated the conversation: the information superhighway.

Bill Clinton's White House was promoting a national vision of a rich and well-funded network that would connect all Americans with one another and the world. He and his vice president, Al Gore, promoted the superhighway imagery in their promises to promote and fund information and telecommunications technology. They depicted the superhighway as a virtually unstoppable stream of information in need of government intervention—especially for cracking down on illegal content and keeping children safe from pornography.[48]

This metaphor was a very useful political and legislative tool, but it did not help Netcom's and Klemesrud's lawyers, who were facing a cutting-edge legal problem: determining whether the companies that help move information move through a network were liable for copyright infringement by their users. This was a major unsolved problem on which there was no congressional guidance until the century's end, when lawmakers passed the Digital Millennium Copyright Act and the Communications Decency Act. (More on these later.)

Without this guidance, Netcom's and Klemesrud's lawyers waged a metaphor battle in the hope that Whyte would develop a view of Internet service providers that released their clients from liability. For example, Randolf J. Rice, one of Netcom's lawyers, told the judge: "It's a fair analogy to say that if you looked at this as a pollution case what the plaintiffs are doing is pointing to a polluted river and suing the river along with the polluter. That is exactly the position of Netcom here. We are simply a means of transmission."[49]

Rice used the river metaphor because it depicted the superhighway as a passive means of transmission. A highway cannot be blamed for illegal materials in people's cars, and a river cannot be blamed for toxic waste that others throw into it. In the same way, Internet service providers could not be blamed for illegal material that users sent through their networks. Rice even argued that making service providers responsible for their users' actions could effectively shut down Internet access. This would be like shutting down the highway because an illegal substance traveled over it.[50]

But metaphors can also encapsulate power relationships, so they can become key points to refute. Andrew H. Wilson, one of the church's lawyers, dug deeper into the highway metaphor and rejected Rice's interpretation of it. He told Whyte: "The analogy is that . . . if you're Netcom you're the owner of a super highway with a lot of other people, and you've got your own little toll booth on an offramp or onramp and you get a call from someone and that person says Mr. Erlich's driving down the road, he's in his Pinto and he's got a trunk full of material, don't let him get on the highway."[51]

Rice did not take the bait. "I would like to work with that analogy," he responded, "but I won't because I think that's always a foolish temptation.[52] Instead, he pointed out that Wilson seemed to be hinting, yet again, that

Internet service providers should monitor all the content they carried in search for copyrighted materials. They were back at Kobrin's technological pipe dream.

Richard A. Horning, Klemesrud's lawyer, added weapons dealers to the list of metaphors Whyte could consider. Emphasizing the ephemeral role Klemesrud had played in the transmission of information, Horning argued that the church was trying to blame a BBS operator for the actions of customers over whom it had no control. The heart of the conflict, he said, was "some form of religious war" in which Klemesrud was an unwilling middleman.[53]

What else was the Internet like? Rice and Horning asked Whyte to think of the suit as a high-tech version of *Smith v. California* (1959), a Supreme Court case on the publication and sale of obscene material. It concerned a bookstore owner in Los Angeles named Eleazar Smith, who had been convicted for violating a city ordinance making it illegal to possess any "obscene or indecent writing," including books.[54] The Court had unanimously sided with Smith, holding, among other things, that the city ordinance violated the First Amendment.

What Rice cared about, however, were not the constitutional details of that landmark case but the fact that the Court had treated Smith's bookstore as a middleman. The same was true of Netcom, he argued: like Smith's bookstore, Netcom was just a distributor that had had "strict liability imposed upon it" and whose only recourse, if RTC had its way, was to find a way of monitoring "everything that crosses the internet in a nanosecond, in the blink of an eye."[55]

This strategy worked. By the end of the hearing, Netcom and Klemesrud were in a much better position than Erlich. Whyte lifted the temporary restraining order against Netcom and Klemesrud and denied RTC's request for a preliminary injunction against them. He ordered RTC to provide a detailed inventory of the items its lawyers had taken from Erlich and issued a modified restraining order that allowed him to keep posting messages critiquing the church. Still, Erlich was explicitly prohibited from digitizing, storing, or transmitting any content for which RTC claimed copyright.

No further seizures could take place without Whyte's explicit approval, but Erlich still had to find a better way to defend himself against RTC's accusations of infringement.[56] And the war would continue: Whyte would see everyone again in a few months, for a full trial.

A REPEAT OFFENDER

Erlich was accused of violating Whyte's order just three days after the hearing.[57] On February 26, he posted transcriptions of excerpts from a tape he had received anonymously in the mail.[58] His post included eight pages out of twenty-five pages' worth of material, alongside nineteen lines drawn from a glossary published by the Church of Scientology. According to Erlich, Kobrin used the incident to tell Whyte that "nothing short of a full ban on Mr. Erlich's postings of Church materials" would "deter him from continuing his violations of RTC's rights."[59]

Erlich took the post down within a day, and he wrote a letter to Whyte explaining that he wouldn't have posted those materials if the court's order had been mailed to his correct mailing address.[60] Kobrin asked for as broad a ban on Erlich's actions as possible, claiming that the incident showed that he was, at best, making a "deliberate decision to remain ignorant of the law" and, at worst, committing "willful violations of the law"—all while claiming ignorance of what was permitted.[61]

As the trial date loomed, the Church of Scientology's attack on online criticism became a media sensation. Wendy Grossman, who was keeping an eye on it all, published her landmark *Wired* article in December 1995, showing just how far the church was willing to go in its raids and lawsuits. Many new targets came to light: a former Scientologist in Arlington, Virginia; a systems operator in Helsinki, Finland; an electronic library and archive in Golden, Colorado; and even the *Washington Post*.[62] Some of these cases involved searches and seizures such as the one Erlich experienced and ended with settlements and court-ordered returns of the seized goods.[63]

Several system operators started reporting their experiences to the Electronic Frontier Foundation, an organization founded in 1990 by three well-known activists: John Perry Barlow, John Gilmore, and Mitch Kapor. The EFF is a digital rights group dedicated to protecting civil liberties online; it funds legal defenses, organizes protests, writes briefs and policy papers, and generally serves as a hub for technologically informed engagement with law, policy, and legislation.

The EFF received several reports that lawyers for RTC and the church were sending copyright infringement notices to BBS operators. In one of its periodicals, the EFF explained that "these threats apparently are designed to convince sysadmins [system administrators] to discontinue the carriage of certain newsgroups that involve discussions of the Church of Scientology in its teachings, solely on the ground that some of the messages sent through these newsgroups allegedly involve infringement of [Church of Scientology] copyrights or other intellectual property rights."[64]

In time, even the EFF received a letter. The church told the foundation that it would not have threatened system administrators with lawsuits if there were any other way to deal with the infringing messages. The EFF deemed it unacceptable for RTC to handle the matter with "the threat of litigation to shut down entire newsgroups," but it also recognized that no culture of conflict resolution existed to handle these Internet-born disputes.[65]

The EFF noted that these unprecedented conflicts showed how important it was "to search for new paradigms" for resolving them. Electronic communications at this scale were still "in their infancy," and service providers such as Klemesrud were "not big corporations with substantial funds to spend on expensive litigation." The EFF suggested that church members respond to online critics by hosting discussions of their own; this would allow them to counter wrongful allegations about them "with more speech." Another option would be to engage in mediation or arbitration, perhaps through an online message board discussion that the EFF could help arrange.[66]

These options would work only if the parties agreed to a ceasefire, but a quick glance at Kobrin's filings and the a.r.s message board was enough to prove that no ceasefire was imminent. The fight was getting very dirty.

Court records suggest that one of RTC's strategies was to call Erlich's moral character into question. Early in the trial, this line of attack had revolved around declarations made by Rosa Erlich Munsey, who had been married to him from 1975 to 1983. Munsey accused Erlich of failure to pay child support, domestic violence, threatening to kill her, and even sexually abusing their daughter. It is unlikely that this testimony changed Whyte's views about the lawsuit.[67]

Munsey's accusations transformed Erlich's case into a polarizing point of discussion on the online forums. Users of a.r.s took sides based on their views of whether Erlich was, indeed, as bad a person as Munsey described. Within a few days, Erlich posted an open letter responding to the accusations, with a focus on his view that the church had compelled Munsey to infuse horrible lies into her telling of their difficult life together. Despite his efforts, the controversy divided what had been a consistent community of supporters, and the damage to his online reputation proved impossible to repair.[68]

Erlich also refused to stop making posts. He no longer uploaded full church documents, but the rest of his posts gave RTC evidence to argue that he meant to continue infringing the church's copyrights unless the court took decisive action. Warren McShane, now RTC's president, found everything he needed simply by following a.r.s for a few days.[69]

On March 10, Erlich had posted: "I have my own religion, thank you. Don't want followers, devotees or paritioners [*sic*]. I do want to be free to preach as I see fit. The US government and the courts thereof have not been granted the right to tell me I can't. Ain't their bizness."[70]

The RTC used this post to argue that Erlich intended to do whatever he wanted regardless of any court order. It was a sequel to a post he made a few weeks earlier, where he had written that the restraining order was "preventing me from doing exactly nothing that I already was not doing in the first place."[71]

Erlich was, of course, free to express himself however he wanted, and out-of-context excerpts hardly proved that he was a continuing threat to RTC. The smoking gun was his response to one user's inquiry. An anonymous user posted a message asking him for an email containing "the illegal file that the s_ts at the Scientology bank are trying to restrict."[72] Erlich responded: "Give me a couple of weeks, till this is over. Then I'll be happy to post it again."[73] In other posts, Erlich referred to a "bogus court order," expressed doubt that RTC actually owned the materials he posted, and used terms that church members considered derogatory.

Defending Erlich would be very difficult. His connection with the EFF helped him reach the national legal spotlight, but any lawyers who chose to defend him would need time to prepare.[74] It would be hard to argue that he had not committed copyright infringement, especially after the latest claims that he had violated the court's order. Several firms turned him down, but after the EFF intervened, a firm known as MoFo (Morrison & Foerster) took the case pro bono.

MoFo was a powerhouse San Francisco firm with offices all around the world. The two MoFo lawyers who agreed to represent Erlich were Harold McElhinny and Carla Oakley. Oakley, a talented midcareer attorney specializing primarily in intellectual property law, had become interested in the case because she found it "unprecedented" and "frightening" that an organization had so quickly obtained a judge's order to "conduct a search of a critic's home." McElhinny was a well-known IP litigator who would later become one of Apple's top lawyers. Like Oakley, he was stunned by the raid. One local newspaper quoted him saying that he "found it frightening" that a federal judge could allow an organization to "conduct a search of a critic's home."[75]

The MoFo lawyers persuaded Whyte to take the case off the calendar for a few weeks so that they could prepare; the judge issued an order on March 17 postponing all hearings.[76] There would be only one more hearing, on June 23, 1995. In the meantime, Oakley countered RTC's multiple attacks with a flurry of legal filings so intense that the church requested that she be formally sanctioned. Whyte himself would write that the court was "disturbed by the parties' seemingly endless applications to the court."[77]

The strategy was to stall and eventually negotiate a settlement. In the meantime, Netcom and Klemesrud would fight their own battle and, it was hoped, provide the MoFo lawyers with ammunition for Erlich's defense. Netcom and Klemesrud's battle was less prominent, but the stakes were higher.

THE GREAT ONLINE SWAP MEET

What would the online world be like if online service providers were liable for users' copyright infringements? If Netcom and Klemesrud lost this battle, they could be required not only to take down all infringing content but to monitor new content for infringement and perhaps even pay financial penalties if anything slipped through the cracks. Nor would this requirement be restricted to Netcom, Klemesrud, and Church of Scientology materials. A victory for the church could have led to an online copyright regime in which service providers had to monitor every bit of content that went through their servers or risk immense financial penalties. This, in turn, would force smaller providers to shut down because they couldn't afford the risk. Whatever providers remained might have to prevent everyday users from posting content altogether.

While the MoFo lawyers handled Erlich's defense, Netcom's and Klemesrud's attorneys were working hard to avoid this situation.[78] They began by moving to dismiss RTC's suit against them.[79] Since their arguments were very similar, I will linger only on Netcom's, which emphasized that the company "leases access to the Internet" by giving users the software necessary to use "the NETCOM gateway."[80] Users could choose one of two pricing options depending on the kind of access they wanted: if they knew some UNIX they could pay $17.50 a month for a shell account; otherwise they could lease a more user-friendly graphic interface for $19.95 a month.

But the company offered access, nothing more. Its brief stated: "NETCOM does not post or control the contents of the information posted or its destination. NETCOM does not control where, when or whether the information may be 'downloaded.' NETCOM does not provide the means for downloading the information. NETCOM, through a fixed-rental arrangement, leases the equipment necessary to enable its subscribers to communicate on the Internet."[81]

This description of Netcom and its users was effective because of the metaphor the lawyers were able to build on top of it: the company leased equipment (software) that allowed users to enter a space (the Internet) and complete a specific task (send and receive messages). In other words, it "merely leases its premises to subscribers who wish to communicate on the Internet and does not in any way select, edit or benefit from what they post."[82]

This framework made Netcom's situation a digital counterpart to *Fonovisa v. Cherry Auction,* an opinion from the Eastern District of California that had recently come to the Ninth Circuit Court of Appeals, which also handled appeals from Whyte's court.[83] This unusual case concerned a swap meet called Cherry Auction. The meet's operators leased space to vendors who traded goods that included counterfeit records and movies. Cherry Auction's staff knew that the vendors were committing copyright infringement, but they did not promote, advertise, or encourage them. After a record label called Fonovisa sued Cherry Auction for infringement, the court found that "merely renting booth space is not 'substantial participation' in the vendors' infringing activities."[84]

This meant that Cherry Auction was not liable for contributory infringement. In suggesting that the same reasoning should apply to their case, Netcom's lawyers were doing something unprecedented: conceptualizing the Internet itself as an intangible swap meet akin to a flea market, thereby bringing into courts of law long-standing metaphors for online message boards (figure 2).[85] In so doing, they bypassed RTC's arguments about the technical possibility to monitor individual communications by arguing that Netcom had no responsibility to do that.

FIGURE 2. A Russian counterfeit cassette tape containing Peter Gabriel's song "So" sits on a counterfeit shelf. If you found this tape at a swap meet, would you want to shut down the entire swap meet to prevent illegal activity, or would you ask the person who brought the tape to leave? This photograph is made available under the Creative Commons CC0 1.0 Universal Public Domain Dedication (https://creativecommons.org/publicdomain/zero/1.0/deed/en). The caption for figure 11 will help you understand what this means.

Whyte took several months to issue his opinions. In September he ordered RTC to returned Erlich's property, but he issued the strict preliminary injunction Kobrin had requested and noted that a full trial on the merits was still pending. He also allowed Erlich to continue posting online commentary but warned him not to reproduce any of the materials covered by the injunction and not to post anything beyond "the extent necessary to carry out his critical purpose." In November, Whyte ruled in Netcom's favor, adopting a very substantial enrichment and expansion of the arguments Netcom's lawyers had proposed. This opinion effectively established that BBS operators are not liable for their users' copyright infringement if they do not actively copy the infringing works or receive financial benefits from them.[86] Erlich and Klemesrud would go on negotiate settlements.

The Internet advocacy landscape was expanding so quickly that civil liberty activists and corporate lawyers had difficulty keeping track of the ongoing conflicts. Religious Technology Center continued to file copyright infringement lawsuits against message board users, generating a collection of complaints so large that some of its critics created websites to archive all the briefs the courts were receiving. One of the greatest changes, however, was that advocacy organizations such as the EFF would soon bring two interrelated communities of advocates together under one roof: those concerned with a recent surge of legislative proposals to create censorship regimes for the Internet, and copyright activists who thought of RTC's copyright suits as a harbinger of new ways of stifling online speech. The stakes were not just creators' control over their own works but users' ability to express themselves and criticize powerful institutions.

Whyte's reasoning illustrates a broader theme: over time, courts and legislators shifted legal responsibility away from Internet service providers. Soon after the district court handed down its opinion on *RTC v. Netcom*, Congress passed the Communications Decency Act (CDA) of 1996, which aimed to regulate online pornography. This law is defunct for the most part, thanks in part to free speech advocates who worked intensively to invalidate it at the courts. However, section 230 still stands, and it reads: "No provider or user of an interactive computer service shall be treated as the publisher or speaker of any information provided by another information content provider."

Section 230 is so important that legal scholar Jeff Kosseff published a wonderful book about it titled *The Twenty-Six Words That Created the Internet*.[87] The section's essence is that ISPs are not liable for the content that users circulate through their networks. It is a broader version of the protection that Netcom was seeking regarding copyright infringement: once users go online, their ISP should not be responsible for their actions. Paired with the safe harbor provisions of the DMCA, section 230 significantly reduced the legal risk involved in letting folks post content online and was arguably essential to the growth of online platforms. For example, imagine trying to launch something like Facebook if you had to risk high fines or a forced shutdown every time a person posts lies about an enemy or a clip from a movie.

But the situation has become very complicated nowadays because online platforms also moderate content, either automatically or through human moderators. Sarah T. Roberts's book, *Behind the Screen,* is a fascinating account of the human labor that this requires.[88] Perhaps you remember hearing about this because content moderation and section 230 were frequent targets of Donald Trump's rage during his first presidency, especially if Twitter removed his tweets or marked them as potentially misleading. In 2020, he signed the "Executive Order on Preventing Online Censorship," asserting that media companies that censor or edit posts aside from harassment and obscenity may forfeit the protection of section 230.

Eliminating section 230, as Trump advocated, would bring us back to the difficult questions from Erlich's time: Could an ISP ensure that none of the content users post is "illegal"? And who gets to decide what counts as illegal? Germany is tackling these questions through its controversial Network Enforcement Act (2017). NetzDG, as this law is known, forces social media platforms with two million users or more to remove "clearly illegal" material (including fake news and hate speech) within twenty-four hours of being posted and to remove all illegal content within seven days. It's too soon to predict the fate of NetzDG, but Erlich's story is a warning: threatening to hold ISPs responsible for user content will meet stiff resistance with the capacity to shake up the conceptual grounding of Internet law. Back in Erlich's time, the ISPs leading the resistance were small, but now Google is leading the fight against NetzDG. It will be a fascinating battle to watch.

CHAPTER TWO

The *Playboy* Suits

ALMOST EVERY TIME I TALK about the Internet, I end up saying something about pornography. Among the reasons for this is that porn is popular, and changes in personal computing and mobile technology have made porn easier to find than ever before.[1] But it was also important in my life as a teenager in Costa Rica, where machismo and homophobia made it nearly impossible for me to ask questions about gay sex. Back then, all my porn was illegally obtained, and it definitely constituted copyright infringement.

Globally, online porn is so popular that technical publications routinely recommend special protocols to ensure that the online traffic it generates doesn't clog a server. A study from 2016 titled "The Internet Is for Porn" analyzed 323 terabytes of online traffic information for eighty million users collected over the course of a week.[2] It concluded that users' engagement with porn sites is so uniquely intense that online servers should account for porn traffic separately in their network resource allocation and traffic forecasting models.

But there is also a legal reason why I keep returning to porn: the industry's lawsuits have been central in delineating what it means to exercise our rights in the digital world. Some of these battles made national headlines. For example, *Playboy*'s victory at the Supreme Court in *United States v. Playboy* (2000) established that section 505 of the Communications Decency Act was unconstitutional. This section required cable companies to scramble or block pornographic channels or to air them only from 10:00 p.m. to 6:00 a.m., when children were unlikely to watch them.[3] It was one of the portions of the CDA that still stood after the landmark case of *Reno v. ACLU* of 1997, in which the Court unanimously ruled that the act's anti-indecency provisions violated the First Amendment guarantee of freedom of speech.

United States v. Playboy was not just about allowing porn to flow freely. It affirmed that a law restricting the flow of information solely because of its content violates our First Amendment right to free speech. As Justice Anthony Kennedy put it at the time: "If a statute regulates speech based on its content, it must be narrowly tailored to promote a compelling Government interest."[4] But the vote at the Court was 5–4. The form our freedom of speech would take online hinged on the opinion of *one* justice. If *Playboy* had lost this case, the consequences wouldn't just be restricted to cable. Children can access the Internet at any time. Perhaps online porn would not have flourished as it did, and more important, perhaps the Communications Decency Act would have left us with something akin to China's Great Firewall.

Researchers have known for a long time that pornography and free speech law go hand in hand. Nadine Strossen's classic work *Defending Pornography* is a wonderful introduction to the topic.[5] However, the relations between pornography and copyright also deserve our attention because they reveal problems and turning points in the broader relations between technology and law. For example, *Playboy*'s legal battles in the late 1990s, which I narrate in this chapter, could have left us with an entirely different online world—one in which it would have been nearly impossible for Internet service providers to transmit images or videos.

In other words, by forcing courts to grapple with the legal implications of how data flows in a computer network, *Playboy* helped shape what it means to have and enforce copyrights online. But this did not happen overnight. It happened slowly over several years, as publishers struggled to create profitable online presences and managers of local networks realized that porn companies would not hesitate to shut them down.

RUSTY N EDIE'S BBS

In 1985, an insurance professional in Ohio named Russell (Russ) Hardenburgh bought a $4,000 Tandy 1200 computer for his small business. The Tandy Corporation, which ran the Radio Shack chain of electronics stores, introduced this device to extend its popular line of home computers into the office market and to offer a cheaper counterpart to the IBM PC.[6]

Hardenburgh added many enhancements to his device, including a ten-megabyte hard drive. He paid extra for the hard drive because the Tandy salesman told him that ten megabytes was all the storage capacity he would ever need. He also splurged on the best color graphics, which he would later describe as "CGA in all its glory, it was beautiful."[7] He quickly learned how to use the software he needed to incorporate the new computer into the business, specifically Tandy's DeskMate package, which included word processing, spreadsheet analysis, and even email and could be integrated with other popular office software. But Tandy was not known for its business applications, and in 1986, Russ splurged on an IBM XT for his office—a cutting-edge PC with a whopping twenty megabytes of memory. The Tandy 1200 came home.

According to his Tandy salesman, the first step in turning his old business machine into a home computer was to purchase a modem. Then, for a $6 copying fee, he could get a program called PC-Talk that would let him connect with other users through a BBS. At the time, Hardenburgh said later, he thought "I could get all the software I wanted free, I couldn't believe it, free . . . from now on free and as much as I wanted, now at last I could have millions of programs, and all free."[8]

Jumping into the BBS world was not easy. He had trouble signing up for an account. "I was told and was advised I should DOWNLOAD a registration form," he recalled. "To be honest it took me 10 days to figure out how to DOWNLOAD, but I finally figured it out and got my form, which I printed and filled out, more questions, I couldn't figure out what my grandparents political party had to do with computers, but I answered every question, I was called and verified about a week later and accepted into the BBS."[9]

Once he had an account, Hardenburgh was frustrated yet again by the very strict rules. In some BBSs, he had to keep a five-to-one ratio—that is, to upload five items for every item he downloaded. He also suspected that the human beings running the BBSs were not very pleasant: he recalled having to "go through the humiliation of various torture routines set up by sadistic SYSOPs, System Operators," which he described as "a very humbling experience."[10]

Still, it was worth it. The BBSs opened up seemingly endless possibilities. He could get anything he wanted: pictures, messages, even programs, for free. He said he was "hooked" on BBSs and became a "program junky." He started downloading everything he could onto his hard drive, which of course soon proved his salesman wrong. Twenty megabytes were far from enough storage.[11]

Hardenburgh became so enamored with BBSs (and frustrated by their rules, staff, and paperwork) that he decided to make his own. He discussed this with his wife, Edie, who had never used a computer. She agreed to the project, and the two of them came up with a name: Rusty n Edie's BBS, which they launched on May 11, 1987. They called their new service "The Friendliest BBS in the World" and distributed posters at computer stores and supermarkets. Folks would just need to call in using their modem, connect to Russ's Tandy 1200, and enjoy themselves. The BBS would "live by the three no's: No Censorship No Rules No Hassle."[12]

The Hardenburghs described themselves as a "couple of burnouts from the 60's" who had never liked rules and hoped to see users "come on in and relax" and feel they were among friends. The couple would not impose time limits or download/upload ratios, and they made themselves personally available to any of their users. The plan was to make money through subscription fees that did not run higher than $10 per month.[13]

The BBS was a labor of love. At first, they hosted fifty files and answered about six calls a day. They were so committed to taking charge of their dream BBS that they even posted their personal telephone numbers so that users could reach them. None of their callers would feel as though a mean person was running the show. Soon it was time for a hardware upgrade. The Tandy stopped working within a few months, and the couple realized they needed more equipment and space. They used the BBS's revenue to transform their basement into a BBS center, eventually acquiring more than a hundred computers and just under fifty modems. These were all networked together and connected to enormous batteries that supplied ten kilovolts of uninterrupted power.[14]

The setup made the house so hot that they had to open their windows, even during the winter, just to keep things cool. In the summer they cooled their basement using a four-ton air conditioning unit designed for a four-thousand-square-foot home. Their living quarters got so dry that they installed two humidifiers. And there was so much static electricity in their home that it felt as if they had "thunderbolts erupting out of our bodies paralyzing one of our nodes or servers." Their cat seemed to enjoy the heat. You can see the cat at the link in the notes.[15]

By 1990, the Hardenburghs and their employees were managing more than sixty thousand files and answering more than three thousand calls a day. They still celebrated their BBS as the friendliest in the world—a digital community space with no censorship, no rules, and no hassle. They continued to advertise that they would answer users' emails and comments, post their own address, and share their voice phone numbers. Users would enjoy extraordinary customer service while navigating a broad range of material: public domain and shareware software, fifteen interactive games, a very active chat and messaging section, over a gigabyte of family-friendly graphics, a dating service, email, and three gigabytes of pornography.[16]

There is some disagreement over just how big Rusty n Edie's BBS became. Court documents and media reports estimate that they had between six thousand and fourteen thousand subscribers by the spring of 1993, each paying $89 a year for access.

By 1990, however, it had become obvious that having no rules or usage limits was untenable. The couple set up a system where for each megabyte that a user uploaded, their download cap would be increased by a megabyte and a half. They also created, or tried to create, a manual system to filter out inappropriate content. It worked like this: Anything that subscribers uploaded would be sent to an "upload file." An employee would check each upload to ensure that the materials were "acceptable" by verifying that the files did not contain child pornography and did not violate copyright law.

This system was in place only symbolically, of course, as Rusty n Edie's celebrated and even advertised the vast amounts of pornography it hosted. The Hardenburghs estimated that they had about 110,000 image files available for download (including about 40,000 pornographic images), but this number was increasing all the time. The files on their servers included a very wide range of computer programs, some of which were illegally obtained. One reporter estimated that about 100,000 of the programs were shareware—software initially distributed for free, with the understanding that the user would buy it if they wanted to keep using it. Otherwise, the software would lock itself after some period or prevent access to key functions.

Still, some of the software had been obtained and posted without their owners' permission, which meant that Rusty n Edie's BBS was distributing pirated software (even if the couple didn't know about it). The Hardenburghs' legal troubles began with a complaint from someone who had made one of these programs: a former restaurant manager in Kansas named Bob Fairburn.

A SELF-TAUGHT PROGRAMMER

Bob Fairburn survived a heart attack in the late 1980s, but his doctors told him he had just five years to live, making him understandably worried about his family's future. Unable to get life insurance, he thought he needed either a miracle or a very profitable new venture if he hoped to guarantee financial stability for his wife and children after his death. He would have to make something that could outlive him and generate a steady supply of cash. According to a popular BBS magazine called *Boardwatch,* Fairburn "decided that there were two things a man could do in America to generate ongoing income—write a book or invent something."[17] He tried writing but was so disappointed by his manuscript that he turned to inventing. If the myth of Silicon Valley promised one thing in the 1990s, it was that a garage and some gumption could make anyone a millionaire.[18]

Fairburn's son was an avid computer gamer. He had a personal computer, had learned how to program in a language called BASIC, and had even started making his own games. He seemed to love his machine and use it with great ease. Fairburn decided that if his son could learn how to make things on a computer, surely he could learn it as well.

Boardwatch's article on Rusty n Edie's BBS, which you can access through the links in the notes, depicts Fairburn as a self-made innovator harmed by BBS culture.[19] He learned how to use the computer, read some books on computer programming, and slowly started writing his own code. Once he was more comfortable with networking, he signed up for EXEC-PC, a very popular BBS run by a Wisconsin operator named Robert Mahoney that for a while was the largest one in the country, with thirty thousand regular callers at its height. From EXEC-PC, Fairburn could download code fragments, tutorials, sample programs, and other self-teaching materials. He became fluent in BASIC and then PASCAL, and soon felt ready to try making a program he could offer for sale.

Fairburn created Home Designer, a program that allowed users to design floor plans for homes and offices. After searching for a publisher for many months, he signed with a Florida company named Expert Software, which renamed the program Expert Home Design and offered it for sale at

$14.95. Bob would earn $.50 in royalties—a little over 3 percent—on each copy sold. Despite these low royalties, however, Fairburn was seeing some profit, and he decided to go into business as a software developer. He purchased a farm outside of Leavenworth, Kansas, hired an assistant, and kept making programs. The plan was to write programs that he could license to software publishers such as Expert Software. This would keep production costs low, even if it meant low royalties.[20]

Now an expert BBS user, he knew he had to remain vigilant in case folks started downloading his software without authorization. This was common knowledge among software publishers, and trade organizations such as the Software Publishers' Alliance routinely ran antipiracy campaigns.[21] A few months after signing with Expert Software, he found Expert Home Design listed in the software download directory of EXEC-PC. He called Robert Mahoney, who reassured him that the program would be taken down and offered an apology.

Users, Mahoney explained, did not always "know the difference between shareware software and commercial software," and they would sometimes upload commercial programs in an effort to keep up with the BBS's upload/download quotas. He added that "most BBS operators will remove it immediately if you call their attention to it."[22]

Fairburn decided to take matters into his own hands. He examined the downloaded file and noticed that it had a note (probably a text file) saying that the file had been downloaded from Rusty n Edie's BBS. He made an account, searched through the BBS, found his program, and immediately sent a message alerting the Hardenburghs. Russ responded through the chat system—"rather rudely," according to the programmer—that he couldn't be held responsible for every file that was uploaded to the BBS.

Fairburn recalled Russ telling him that Rusty n Edie's received many megabytes of content each day and that he would remove the file when he had time. A week passed and the program was still there. Remembering this moment a few months later, Fairburn told *Boardwatch Magazine,* "Understand . . . I'm not Bill Gates. I only get fifty cents per copy sold, and my family depends on this for a living. This guy was running a giant bulletin board and taking in lots of subscriptions, and basically he was stealing my software. I just got mad about it."[23]

Frustrated and angry, Fairburn contacted the Kansas City office of the Federal Bureau of Investigation, which sent an agent to his home. The visit was amicable but discouraging, as Bob realized very quickly that the agent did not know much about computers. They logged into Rusty n Edie's and searched through the software listings to find Expert Home Design, Quicken, several Microsoft programs, and "virtually every commercial game program" ever made. Even if the agent couldn't use computers, he couldn't deny that the programs were there.[24]

Fairburn gave the agent a disk with a log of the session and a few sample files they had downloaded. He also added a copy of PKZIP—the compression software he had used to make the files fit into the disk, which the FBI agent needed to learn how to use. Over the next few months, the FBI contacted the publishers of many of the other programs they had found and told them that the programs were available at Rusty n Edie's. The FBI prepared to raid the Hardenburghs.

Being raided by the FBI was not an unusual occurrence in the hacking scene of the 1990s. Bruce Sterling wrote about this in his classic book *The Hacker Crackdown:* law enforcement officials' inability to police online spaces pushed them to intervene at the physical spaces where hackers lived and worked. In fact, while the FBI was preparing to raid the Hardenburghs, computer magazines were following a raid that culminated with a game company suing the Secret Service for invasion of privacy and scoring a partial win in federal court.[25]

On January 30, 1993, the FBI arrived at the Hardenburghs' home and raided their basement. They took 130 personal computers and many modems, cables, software packages, and subscriber records. Rusty n Edie's—one of the country's largest BBSs and, according to a competitor, "the worst-kept secret in the industry"—went offline. Russ Hardenburgh told *Boardwatch* that he "never thought that this could happen in America" and vowed to restore the system as soon as possible. But Russ's legal problems were only beginning.[26] What was unusual about this raid was not that it happened to a small BBS but that the materials the FBI collected would give *Playboy* the evidence it needed to shut the Hardenburghs down.

PLAYBOY ONLINE
The cover of *Playboy* in February 1985 featured a special section on women from Texas (to quote the cover: "Eat Your Heart Out, California!") and an interview with Steve Jobs, the "29-year-old Zillionaire" from California. The surprisingly thorough interview, nearly thirteen pages long, covered

everything from Jobs's youth to his plans to overtake IBM in the business machine market.[27]

It isn't surprising that *Playboy*'s editors were interested in Jobs. According to the historian Elizabeth Fraterrigo, the magazine's rise to fame starting in the 1950s revolved around a vision of "the good life" that directly challenged the family-centered ideal embraced by postwar America. In the view of *Playboy*'s publisher, Hugh Hefner, men should surround themselves with high-end goods and beautiful, sexually available (primarily white) women.[28] Jobs had a very marketable story for *Playboy*'s readers: a twenty-nine-year-old looking confident in suspenders and bowtie, sitting on almost half a billion dollars, and ready to take on a Goliath-sized competitor, IBM.

But interesting interviews are not what made *Playboy* famous. Its first issue, in 1953, gained immediate notoriety by featuring a nude photograph of Marilyn Monroe. In the 1970s it became the first mass-market magazine to feature female frontal nudity. By one estimate, in the mid-1970s more than a quarter of all US male college students bought the magazine each month. Jobs's interview landed squarely on a demographic intersection: the male college students and graduates whom *Playboy* reached by the millions overlapped substantially with the groups Jobs hoped to reach with his latest computer, the Macintosh.

In 1987, the alliance between Apple and *Playboy* led to the magazine's first foray into cyberspace: a newsletter called *Playboy Online*. It was available exclusively to Macintosh users with subscriptions to a small group of popular online service providers and BBSs. It featured articles, interviews, and the playmate of the month. Both companies would benefit from the arrangement: businessmen would have an additional incentive to buy a Mac, and *Playboy* readers would have an official place they could go to for porn (as opposed to searching for *Playboy* images on third-party message boards).[29]

Playboy Online was meant to encourage people to buy print copies. Eileen Kent, a contracts administrator at the magazine, described *Playboy Online* as the first publication, other than computing industry newsletters, to feature a combination of high-quality graphics and texts. She told a journalist: "We hope Macintosh users who see Playboy Online will enjoy it enough to go out and buy a copy of the magazine, or become a subscriber."[30]

The newsletter was also meant to test whether it was possible to create a vibrant online space for authorized dissemination of *Playboy*'s images. Paul

McGraw, the digital design firm in Missouri that produced *Playboy Online*, thought of the newsletter as an experiment in translating print standards to online settings. He said, "I think everyone will be impressed by how we're able to take to a print publication with high production standards and make it available online."[31]

The test failed. *Playboy Online* was shut down just five months after its launch. The trade literature for the publishing industry suggests that it was killed by a mix of high operating costs and the economics of early online advertising, which was already becoming a driving economic force behind online communication. *Playboy* had tried to recover its operating costs by selling advertisements in the newsletter, but connecting to a BBS was expensive because it involved a timed phone call and the upload/download quotas that the Hardenburghs hated so much. This meant that readers of *Playboy Online* did not want to spend precious connection time downloading advertisements in order to see images they could get from other sources.

In 1988, Christie Hefner, Hugh's daughter, took over the magazine. Under her leadership, the company would continue to expand its presence across the rapidly developing multimedia landscape. Her crowning achievement in this field was the launch of playboy.com in 1994, coinciding with the spread of early World Wide Web browsers such as Netscape.com. In the half decade between the demise of *Playboy Online* and the launch of playboy.com, the company worked very hard to ensure that *Playboy* magazine images did not circulate online without the company's consent. This involved aggressive litigation and a new line of work: finding and cataloging online porn.

This unusual job was given to Anne Steinfeldt. Every day she would go online, catalog any *Playboy* content she found, and report back to the company. She monitored the BBS landscape, regularly checking popular ones and keeping an eye on those that seemed to be growing. One day in 1992, she logged onto Rusty n Edie's under the pseudonym "Bob Campbell," did some keyword searches, and found about a hundred erotic GIFs that appeared to be *Playboy* content. She saved all these images into floppy disks and delivered them to Timothy Hawkins, Playboy's photo librarian, who searched through his library's collections and confirmed that Playboy owned those images. Her work downloading and classifying images was essential to Playboy's legal defenses and online presence. It is an example of what media scholar Dylan Mulvin calls "infrastructural labor" in his fabulous book, *Proxies*.[32]

Armed with this evidence, Playboy filed a complaint against the Hardenburghs at the District Court for the Northern District of Ohio on March 11, 1993. The company moved for summary judgment—one without a full trial—for copyright infringement of ninety-nine images that Steinfeldt and Hawkins had reviewed. Playboy alleged that the Hardenburghs had committed both direct infringement (copying the original elements of a copyrighted work) and contributory infringement (knowingly contributing to or assisting another party's infringing activity).

Playboy hired a large full-service law firm called Porter, Wright, Morris & Arthur and an IP firm with extensive experience in information technology, Klarquist, Sparkman, Campbell, Leigh & Whinston. The Hardenburghs hired the Ohio defense firm of McNeal, Schick, Archibald & Biro. A Silicon Valley attorney named Thomas O'Donnell was working with them. One striking aspect of this lawsuit is that Playboy received access to the materials the FBI had collected when its agents raided the Hardenburghs' basement—even though that raid was conducted under unrelated accusations of software piracy. Playboy's lawyers examined the FBI's materials and claimed that twenty more of its images were in the Hardenburghs' servers. The Hardenburghs countered that the images had been placed there by their users, not themselves.

Playboy's lawsuit moved slowly through the court, in part because of the death of one of the judges hearing the case. This additional time allowed Playboy to scan the FBI's archive more thoroughly; its lawyers found an additional 392 GIFs that they could trace back to the magazine. They assembled thorough records identifying each GIF, the certificate of copyright registration for the original photograph, and a photograph from the magazine's libraries.

INTANGIBLE DISPLAYS

Since it was clear that the Hardenburghs' BBS was hosting and distributing quite a lot of *Playboy* material, the main issue for the court was whether Russ and Edie should be held responsible for their users' infringement. The court followed Playboy's lead in adopting a theory of online liability grounded on both the magazine's earlier lawsuits and some recent developments in video game copyright. For this reason, I find myself less interested in the details of how the trial against the Hardenburghs unfolded than in one of the biggest questions that it raised: What does it mean to "display" something?

FIGURE 3. A framed photograph of my parents I keep in my office. This picture's journey from my disposable camera, through a scanner and several hard drives, and into my office computer and printer, illustrates the many meanings that the term "display" could take in the 1990s. I took this photograph.

This is a surprisingly difficult question. Take a look at the framed photograph of my parents (figure 3). I took it fifteen years ago using a disposable Kodak camera. After the Kodak store gave me the printed photograph, I scanned it and stored it in an external hard drive, and I took that hard drive with me when my husband and I moved to California. Six years ago I found the image on the hard drive. So I opened my Gmail account, attached the image to a message, and sent the message to my UC Davis email account. Once I got to my office, I printed the image and left it on my desk for a few weeks while I bought a frame. Eventually I put the image in the frame and placed the frame on my desk.

At what points in this journey could we say that the image is "displayed"? We can all agree that the frame on my desk counts as a display: the image is placed in a location that allows me and my visitors to see it. But was it being displayed when I left the printout on my desk for a few days before framing it? Students and colleagues visited me, and they surely saw the image. Perhaps I displayed an image without deliberately creating *a* display, and the act of displaying can happen without my intention. I kept the image on my desk not so that people would see it but because I delayed getting a frame.

Now let's think about the image's digital journey. At what points between scanning and printing was the image displayed? I was working alone, so if any point in that journey counts as a display, then we must abandon the assumption that a display requires an audience beyond the person manipulating the file. The moments when a screen shows a digital version of the image are the top candidates for what it would mean to *display* the image. Those are the only moments when human beings would be able perceive it. Such a display could range from a thumbnail icon to a high-resolution version, though, as we will see in chapter 8, some ways of showing the image would not count as a display in the twenty-first century.

Playboy's lawsuits counted an image's presence on a screen as a display, but they did not stop there. Instead, they aimed to establish that an image's mere presence in a device counts as a display. This meant that whether a human being perceived the image was a sufficient, but not necessary, condition to claim that a display had taken place. Simply by *having* the file on a computer or transmitting it through a network, a user or Internet service provider was displaying the image. If we apply this logic to the photograph of my parents, everything that happened to the image, starting from the moment I scanned it, was a display—even the data exchange that moved the image from my Gmail address to my UC Davis account.

The reason Playboy's lawyers pursued this expansive definition of a display is that copyright law establishes a display right. The Copyright Act of

1976, which codified decades' worth of common law, established that a right "to display the copyrighted work publicly" applies "in the case of literary, musical, dramatic, and choreographic works, pantomimes, and pictorial, graphic, or sculptural works, including the individual images of a motion picture or other audiovisual work."[33]

But establishing that a digital image has been displayed was difficult because of the distinct materiality embedded into the Copyright Act. It defined "copies" as "material objects, other than phonorecords, in which a work is fixed by any method . . . and from which the work can be perceived, reproduced, or otherwise communicated, either directly or with the aid of a machine or device." This would then inform the act's definition of "to display," which was "to show a copy of it, either directly or in directly or by means of a film, slide, television image, or any other device or process." In other words, to display, according to the literal text of the act, was to show something tangible.[34]

Playboy's legal goal in the 1990s was to secure an interpretation of the act that established rights over intangible displays. In its fight to do this against the Hardenburghs, the magazine's greatest weapon was one of its own lawsuits, *Playboy v. Frena*. This case, decided at the Middle District of Florida in 1993, concerned a system operator for Techs Warehouse named George Frena. Techs Warehouse was a BBS in Florida that—like Rusty n Edie—carried a vast content library that included pornography.

Frena offered free access to anyone who purchased goods from him through the BBS and otherwise charged $25 per month. Unlike Rusty n Edie, though, Frena did not have a person moving content from an "upload" to a "download" folder. Instead, whatever subscribers uploaded was immediately available for download. Frena insisted that he had not uploaded any of the pictures himself, but he did concede that his BBS contained at least 170 unauthorized *Playboy* and *Playgirl* images.

The Florida court followed Playboy's logic to find that by providing a "space" for photographs to be uploaded and downloaded, Frena had violated the magazine's exclusive right to distribute its images. It didn't matter whether he made the copies himself—he had "supplied a product containing unauthorized copies of a copyrighted work."[35] In this reading of the law, having an image inside a server was like having a framed photograph on a desk. It didn't matter whether other people were looking at it: if the frame sits on a desk, it is on display.

Yet the materiality that the act embedded into its definition of "copy" and "display" did not match what key lawmakers understood as their intent. In the 1970s, when Congress was debating the bill that would become the Copyright Act of 1976, lawmakers deferred discussions of information technology to a separate group called the National Commission on New Technological Uses of Copyrighted Works.[36] This was meant to prevent further delays in the passage of new legislation, as there was growing industrial and legal pressure to align the US copyright system with international treaties.

The commission is best known today for recommending what eventually became the Computer Software Copyright Act of 1980, which established computer programs as a new category of creative work eligible for copyright protection. While the commission was hard at work, however, Robert Kastenmeyer (a Democratic representative from Wisconsin and a key sponsor of the bill) entered his intent regarding intangible displays into the *Congressional Record*. He submitted a report to the House of Representatives conveying copyright owners' demands for a definition of "display" broad enough to encompass digital transmission. Specifically, a "display" would include "the projection of an image on a screen or other surface by any method, the transmission of an image by electronic or other means, and the showing of an image on a cathode ray tube, or similar viewing apparatus connected with any sort of information storage and retrieval system."[37]

In Playboy's hands, this became a crucial legal theory: despite the materiality built into the Copyright Act, Congress intended that the definition of "display" include transmission of an image through a computer system. This blurred the lines between "display" and "distribution" and provided the grounding for the court to rule that Frena's actions had violated two exclusive rights. This violation, along with evidence that Frena had access to the copyrighted works, was enough for Playboy to convince the court that he had committed direct copyright infringement.

Lawsuits unfolding in the video game industry added weight to this legal theory of intangible displays. For instance, in 1994, the District Court for the Northern District of California decided *Sega Enterprises v. MAPHIA*.[38] Sega, the makers of popular games such as Sonic the Hedgehog, sued a BBS for offering twelve Sega games, ten Sega-licensed games, and six games that the company was developing, including soon-to-be classics such as Jurassic Park and Sonic Pinball.

The court found that Sega had established that "unauthorized copies of its games are made when such games are uploaded to the MAPHIA bulletin board" and "when they are downloaded to make additional copies by users." The court also cited *Frena* to rule that even if the defendants "do not know exactly when games will be uploaded to or downloaded from the MAPHIA bulletin board, their role in the copying, including provision of facilities, direction, knowledge and encouragement, amounts to contributory copyright infringement."[39]

The *Frena* rationale, especially as the *Sega* court interpreted it, made a clear impact on copyright law for three reasons: it detached the concept of the "copy" from the materiality built into it in the Copyright Act; it established computer transmission as an act of display and distribution; and it allowed copyright owners to pin copyright infringement on someone who allowed an unauthorized image to be stored in or transmitted by a server. In short, as far as online servers were concerned, *having* or *transmitting* an image was the same as *displaying* it.

Unlike Playboy Enterprises, the Hardenburghs did not have much legal precedent on their side. Their defense was grounded on the notion that it was the BBS's subscribers, not its staff, who uploaded and downloaded *Playboy*'s images. Certainly, Rusty n Edie's provided an incentive for members to do so, but the subscribers were the ones who had scanned the photographs, made digital files, and uploaded them into the system. The Hardenburghs' lawyers argued that Rusty n Edie's served as "passive providers of the space in which the pictures were passed from one party to another."[40]

This argument would have allowed them to build on the reasoning of *RTC v. Netcom*, the Scientology case discussed in chapter 1. Back then, Judge Whyte had been especially critical of *Frena*. He had written that a *Frena*-style ruling for the Church of Scientology "would result in liability for every single Usenet server in the worldwide link of computers transmitting [a] message to every other computer." With that reasoning, Whyte had concluded that simply creating "a space where infringing activity occurred" was not grounds for infringement liability.[41]

For this to work, though, the Hardenburghs needed to argue that their BBS, like the one in the Scientology case, resembled a swap meet. Swap meets were the guiding metaphors on which the Scientology cases were decided because they allowed for direct comparisons with the *Fonovisa* opinion. And there was the crack in the Hardenburghs' defense: since someone on their staff moved files from the "upload" folder to the "download" folder, they were unable to argue that they couldn't create a system of human oversight. Even if the system was imperfect, it showed that it was possible to surveil the content the Hardenburghs hosted.

The Hardenburghs lost on all major fronts. In 1997, Judge Sam H. Bell ruled that the couple had displayed the content; that the displays had been public (that is, that they occurred outside normal circles of family and social acquaintances); and that the Hardenburghs had committed direct infringement. Bell also agreed that the reasoning in *Sega* was applicable to Rusty n Edie's BBS, ruling that "the mere creation of a BBS is sufficient to establish direct infringement liability where copyrighted material *appears* on the system."[42]

Putting all this together, Bell granted summary judgment to Playboy on its claims of direct and contributory infringement. The Hardenburghs settled the case in 1998, but they soon started seeing other network service providers win battles that resembled their own. If only Playboy had sued them a few years later, perhaps they would have won.

©©©

The Digital Millennium Copyright Act, passed in 1998, transferred much liability for copyright infringement from Internet service providers such as the Hardenburghs to the users of those services. Title II of the DMCA is called the Online Copyright Infringement Liability Limitation Act. It establishes something called "safe harbor provisions," which specify that an ISP is protected from copyright infringement liability as long as they adhere to certain rules. The most important one is that they must have a takedown procedure such as the one I mentioned earlier. They must also not have benefitted financially from infringing activity and be able to prove that they did not know that they were hosting the copyrighted materials.

I could not stop thinking about the DMCA when I was writing up the Hardenburghs' story, so I decided to investigate what porn companies' copyright complaints against users were like. My research led me to a few people who have settled with these companies. However, telling me even one detail of the terms of their settlement could open them up to breaches of settlement and financial ruin, so I did not feel comfortable pursuing personal inquiries as a line of research. After all, if there is one thing I have learned from the social studies of privacy, it is that it can be very easy to deanonymize information.

My way around this problem was to dive deep into porn sites and message boards in search of information. I noticed several patterns that I can merge into a plausible composite picture that matches the instincts I developed while writing this book. To protect myself and the authors of the sources that I found, I will not name individual websites or quote any posts directly in the explanations that follow.

There are countless kinds of pornography repositories online. Some of them are search engines. In response to a search term, the website generates a grid of thumbnails of video screenshots. Clicking on any of them will direct the user to another site where the content is hosted. Other websites have rules that resemble BBS users. Users create an account tied to their email. This gives them access to a wealth of pornography that other users have uploaded, but they must maintain an upload/download quota if they wish to keep their membership to the site. Some of the sites host the material, but others simply require users to place their file in a peer-to-peer network such as BitTorrent. (We'll discuss this kind of network in the next chapter.)

These exchange-based websites are the ones that interested me most when thinking about the Hardenburghs' story, so I followed users and videos across sites and social media looking for any legal trouble tied to uploading. I found dozens of online posts about copyright complaints filed against these sites' users. Some folks learned about this indirectly when the company contacted their ISPs to request their contact information. Others heard from the company directly when their accounts were easy to deanonymize— perhaps because they used their personal email address or linked their account to a social media platform. The companies were uniformly willing to settle, though the amount required to do so varied wildly. Some users reported fines of a few hundred dollars, and the more alarming settlement amounts were calculated by the number of times the unauthorized video had been viewed or uploaded. In those cases, the settlement amounts could jump to hundreds of thousands of dollars overnight if a video went viral.

These findings are consistent with ongoing revelations of the extreme litigation strategies that some pornography companies employ. For example, in 2019, Bloomberg.com published a piece titled "Porn Purveyors' Use of Copyright Lawsuits Has Judges Seeing Red."[43] It explained that two porn vendors had filed more than 1,700 copyright infringement lawsuits in the first seven months of the year. That's more than half of the total number of such lawsuits filed. In fact, such lawsuits seem to be so frequent that some law firms advertise defensive services. The website for the Russell Firm asks, "Received a Strike 3 Holdings Subpoena Letter? Let Me Defend You." Another firm, Antonelli Law, advertises: "You may feel the aim of Strike 3 Holdings is to shock and embarrass you into paying a substantial settlement on their terms. We can help you figure out the best course of action for your particular case."[44]

Let's pause and reflect on the difficult situation that this generates. Certainly, creators have a right to uphold their copyrights. But the reality is that determining who the targeted user should be and determining the severity of their transgression can be very difficult. Think of a tech-savvy tween exploring their sexuality for the first time, realizing that they can get videos from the one performer that sparked something in them by sharing videos that they already have. Consider how easy it would be to create a fake account in someone's name and upload materials in their name or, conversely, to receive a threatening notice from an ISP while having done nothing wrong—just because a house guest happened to upload or download something while visiting you. This is about much more than pornography. It is about what it means to share content online and about the DMCA's impact on people's everyday interactions with the web.

CHAPTER THREE

The Napster Hydra

I'VE ALWAYS BEEN A COLLECTOR. When I was a teenager, in the months leading up to Y2K, my collections started to take a digital form. My parents had bought an enormous Hewlett Packard desktop computer and installed a separate phone line for me to connect to the Internet. I still remember the first website I visited (the homepage for E! Entertainment Television) and my first foray into pornography (playboy.com, which helped me confirm I'm gay).

Every day after school, I saved dozens of photographs to my hard drive, categorizing them into folders according to their content: specific television shows or movies, artists I admired, poems that moved me (I was a very moody teenager), and assorted illustrations for all my favorite movies and television shows. I had folders nested within folders, hiding files deep inside folder paths only I could remember. The deepest of these paths led to a treasure trove of pornography—tons of it.

Maintaining my file collection enabled me to express myself in digital form. If I was feeling sad, I would find personal websites where people posted sad poems. Poems that seemed to resonate with my state of mind would go into one of my folders. If I was feeling joyful, I would find glittery GIFs of my favorite cartoon characters and save them near the top layer of my nested folder collection. Every now and then, I would share some of these things onto my own website on the hosting service Geocities.

I mainly collected text and images until, early in 2000, I heard about a program called Napster. It sounded like magic. The website explained: "Imagine . . . an application that takes the hassle out of searching for MP3s. No more broken links, no more slow downloads, and no more busy, disorganized FTP sites. With Napster, you can locate and download your favorite music in MP3 format from one convenient, easy-to-use interface."[1]

Napster connected users directly to each other and allowed them to share files. Unlike traditional downloading, in which a user obtained something directly from a server, Napster allowed users to obtain files by downloading portions of a file from several people who already had it in their computers. That's the heart of the peer-to-peer (P2P) model: decentralizing the file's location and allowing users to download bits and pieces from each other. According to the website, *Napster is music at Internet speed.*[2]

I could not resist P2P sharing. Napster's MP3 files were much easier to download than WAV files, and I didn't need special software to play them: Napster had a built-in MP3 player. In the space of around three megabytes (enough for about five pictures), I could have a song on my computer. I could burn that song onto a CD and play it on my bulky home stereo or the cutting-edge CD player I bought at a used electronics store in Panama. Copyright was irrelevant to me.

Looking at Napster's website from 2000, I recently noticed a subtle but important detail: the site didn't say that I could download any music I wanted (even though that was clearly its purpose). Instead, it presented Napster as a service for artists who wanted to reach a wider audience. "Are you an artist who wants in on the revolution?" the website asked, "Click here to get your music heard on the world's largest online music community."[3]

The point of this pitch was to show that Napster was not *necessarily* meant to be used in a way that infringed on anyone's copyrights. This is an essential legal maneuver that allowed its lawyers to rely on the Supreme Court case *Sony v. Universal,* decided in 1984.[4] (I'll say more about that landmark case and its relation to Napster later in this chapter.)

Napster's story shows how the values and assumptions that guide the development of a new technology can bring about the technology's downfall. It is not that Napster itself was inherently good or bad. But copyright law's effect on the early online music scene would come to reveal the conditions under which a new technology is, in itself, illegal.

THE DMCA AND THE RECORDING INDUSTRY

My media collection was one of millions around the world. Pictures, music, books, even software: if it could be uploaded, it could be downloaded. I even had a program called Hacker's Little Helper that helped break some of the rules. In my previous book I described how software publishers in the 1980s lobbied very hard to strengthen intellectual property laws to curb software piracy. Hackers even set up international hacked software shops from the comfort of their own homes. Kevin Driscoll's book *The Modem World* tells how pirated media circulated freely over BBSs well before the development of the World Wide Web.[5]

Works such as these, and many others, detail the great lengths to which software publishers were willing to go to prevent piracy. Their lawyers made garments and swag to celebrate their involvement in antipiracy efforts. Publishers even found ways of making physical barriers against piracy that were fun to play with. Lucasfilm, the production company that made *Star Wars*, had a particularly curious one in the early 1990s: a codex made up of two circles connected at the center (figure 4). Midway through a game, players had to align the images in the circles to reveal numbers and codes they needed to input into the game. The codices were fun souvenirs, and circulating copies of a game without them was useless.

But one of the most pressing needs of software and media publishers around the world was an overhaul of copyright law that would protect their works from unauthorized digital distribution. The World Intellectual Property Organization (WIPO) played a pivotal role in this process. In 1996, with industrial support, WIPO produced two documents known as the "WIPO Internet Treaties." The first of these was the WIPO Copyright Treaty, an agreement to protect works (especially computer programs and databases) in digital environments.[6] It is now a special agreement under the Berne Convention, an international group of countries (including the United States) that follow mutually compatible copyright guidelines.

The second one is the WIPO Performance and Phonograms Treaty, which affirms the digital rights of performers and producers who make sound recordings or live performances. Among the rights it affirmed was the "right of making available," which is essentially the right to post the recording online. Officially, it is "the exclusive right of authorizing the making available to the public of their performances fixed in phonograms,

FIGURE 4. These are codexes that players needed to play Lucasfilm games. Each one consisted of two cardboard disks attached together at the center. The player could rotate the inner disk to align the characters' heads and bodies. As the smaller disk rotates, the punched rectangles on its inside revealed different codes printed on the bigger disk. I took this photograph.

by wire or wireless means, in such a way that members of the public may access them from a place and at a time individually chosen by them."[7]

Ratification of the WIPO Internet treaties pushed the wealthier Berne Convention members to update their own copyright laws to align themselves with the new agreements. Industry representatives from across the media and software landscapes lobbied hard to encode anticircumvention provisions into their national laws. At the European Union, for example, the Parliament in 2001 passed a directive on "the harmonization of certain aspects of copyright and related rights in the information society." The directive followed the WIPO treaties and included a section on "Obligations as to technological measures" that required member states to "provide adequate legal protection against the circumvention of any effective technological measures."[8]

The US Congress responded to the WIPO treaties by passing the DMCA in 1998. The DMCA did several things, but four of them are particularly important to our stories: (1) it formally extended the reach of copyright law into online spaces; (2) it prohibited hacking one's way past technical copyright protections; (3) it mandated the creation of takedown procedures for creators who found their work illegally copied online; and (4) it limited Internet service providers' liability for their users' copyright infringements.

The third and fourth features are part of what lawyers call the DMCA's safe harbor provisions. They protect online service providers who transmit, cache, or link to infringing material. To qualify for this safe harbor, an online service provider must not be receiving a direct financial benefit from the infringing material, must be able to show that it wasn't aware of the material or of anything that implied its existence, and must have a notice and takedown procedure.[9]

The safe harbor provisions were both a lobbying victory and a way of ensuring that the DMCA aligned with the WIPO treaties' directive to immunize online service providers from copyright infringement liability. Had the DMCA been in place in the early 1990s, the main goal of Playboy's suit against Rusty n Edie's BBS might not have been to shut it down but to force the BBS to identify the users who were posting *Playboy*'s images onto the system—and maybe those who downloaded them, too. And the Church of Scientology would not have been able to go after support.com for copyright infringement just because it hosted Dennis Erlich's materials. Their primary copyright infringement battle would have been against Erlich himself.

A political powerhouse named Hilary Rosen led the music industry's early reliance on the DMCA. Rosen had joined the Recording Industry Association of America—RIAA, the main trade organization for the US recording industry—in the late 1980s, and she steadily climbed through the organization's ranks until she became its CEO in 1998. A Democrat who supported many progressive causes, Rosen also was such a substantial fundraiser for LGBT causes that the *Washington Post* called her a "check-writing champion of gay rights."[10] This was a passion she shared with her former wife, Elizabeth Birch, who was the executive director of the Human Rights Campaign from 1995 until 2004.

Rosen was a major presence in Washington, DC, during the late 1990s. Her fundraisers attracted A-list celebrities, intellectuals, and politicians, and she sometimes brought top musicians to the capital to lobby for the music industry's interests.[11] Rosen is not a lawyer; she developed her lobbying expertise in the early 1980s by working for Liz Robbins Associates, a lobbying firm that represented powerful organizations such as Goldman Sachs, the City of New York, and the City of San Francisco. In 1987, when she was twenty-eight years old, Rosen joined RIAA as the group's first director of government relations.

Rosen's record at the Recording Industry Association of America is stunning. She updated the Parental Advisory label program, opposed efforts to censor music lyrics, lobbied strongly in favor of the passage of the DMCA, and celebrated RIAA's issuance of its first Diamond Certifications—an honor it would bestow annually to albums that sold more than ten million copies.[12]

Rosen and RIAA believed that piracy was mainly to blame for the industry's economic woes. The Recording Industry Association's largest unit was its antipiracy division, with dozens of members, most of whom worked on traditional media. The division teamed up with local police agencies to stop the production of counterfeit records; lobbied state legislatures to strengthen anticounterfeit rules; and filed lawsuits against companies that used unauthorized music.

Starting in the late 1990s, however, Rosen and the antipiracy division had to develop a legal strategy to cope with an entirely new situation: the rise of websites and online services that offered free unauthorized music to anyone with an Internet connection. Dealing with this problem would require a careful long-term strategy. On one hand, free and unauthorized access to copyright music could place the music industry in a precarious, even devastating, situation. On the other hand, retailers such as Tower Records were riding the dot-com wave and discovering that e-commerce offered a viable alternative or supplement to traditional brick-and-mortar stores.

Rosen's appointment as the association's CEO in 1998 made her the chief advocate for the music industry, which included almost 350 companies. By then, the industry's explosive growth in the 1980s had slowed markedly, and record companies and retailers were struggling. Even giants such as EMI Records Group had downsized substantially, and recovery was not in sight.[13] In her new role, Rosen developed a reputation for expertise in virtually every aspect of the industry and for seemingly infinite energy and passion for the music industry and progressive causes. "She jets from Washington to Nashville to Los Angeles to Tokyo and home again in constant pursuit of moneymaking opportunities for the music moguls," wrote one reporter in the late 1990s, "She jumps, allegro, from subject to subject, from the technical specifications of DVDs to the musical merits of her friend Sheryl Crow to her pet project, Rock the Vote."[14]

Rosen juggled a vast array of projects, including First Amendment protections for rap lyrics (an issue we'll discuss later) and revising RIAA's copyright efforts in response to the DMCA's passage. The legal and technical landscape for her battles against music piracy were changing simultaneously. For years, the Internet had facilitated the free exchange of copyrighted material. Users could share images, texts, and videos by uploading them to central servers from which others could download them. Central server models had become especially popular for commercial Internet services, in part because most home computers lacked the processing capacity to handle repeated requests from other devices. This is still the norm today.

By the late 1990s, a different kind of connection model had become possible. Home computers were increasing in processing power, networks had higher bandwidths, and compression-based formats such as MP3 and MPEG reduced the file sizes for things such as songs and photographs.[15] This allowed users' own computers to be directly connected to each other, so that one user could download another's files without going through a central server. When I was building my music collection with Napster, I could click on a user's name to see the songs stored in *their* computer. Then, provided that their computer was on and the Napster app was running, I could download them directly.

This new wave of P2P sharing, brought into the public eye by services such as Napster (which launched in 1999), would cause high-stakes legal battles among users, publishers, and developers. Certainly the enormous losses that Rosen's member companies faced were a major concern. The issue, however, was that people's relationship with music itself was changing, perhaps permanently. Instead of having to buy a CD, folks could now download one song. And the first stop to finding a song they wanted to hear was no longer the record shop or a radio station: it was a computer owned by someone they had never met.

NAPSTER

At the MTV Video Music Awards in 2000, Shawn Fanning, a nineteen-year-old dropout from Northeastern University, stepped onto the stage to uproarious applause. Wearing a black T-shirt with a Metallica logo, he strutted toward the show's host, Carson Daly, to the beat of the band's song "For Whom the Bell Tolls." Daly told the audience: "In the last year, this teenager has the developed the technology that has revolutionized the way we get our music."[16]

As millions of viewers around the globe watched, the two men basked in the crowd's attention while discussing Fanning's T-shirt. "A friend of mine shared it with me," Fanning told Daly. "I'm thinking of getting my own, though." The camera panned to Lars Ulrich, Metallica's drummer, who closed his eyes and appeared to doze off while Fanning and Daly prepared to introduce the next performer—pop sensation Britney Spears—who had come, Fanning said, to "sing a song older than she is."[17]

Fanning had become a global celebrity in just months for his development of Napster, the free P2P service that allowed users to share files with each other at record speeds. He had written the program on his Dell notebook computer over the course of several all-night coding sessions, often at his uncle's home. Soon thereafter, he had launched the program in partnership with Sean Parker, a skilled programmer and budding entrepreneur who would later serve as Facebook's first president before moving on to create Spotify.[18]

Napster's early websites were simple and direct: "It's Here!!," announced a landing page in October 1999: "Download it now and join the internet music revolution." The program's main advantage over central server distribution was, of course, its P2P architecture. "Welcome to Napster, the future of music," the site explained. "Napster is the best search engine available, and the best way for users to find and download MP3s. By creating a virtual community, Napster ensures a vast collection of MP3s for download. Napster also eliminates the problems of conventional FTP transferring by using cutting-edge technology to ensure the completion of each MP3 transfer."[19]

Napster offered a broad range of search options. Users could filter results by quality of the sound, and the results page would help users select the files with the shortest download times. Users could also chat with each other on genre-specific forms, and they did not need to download a separate MP3 player to listen to their files. Users could even mark others' song libraries as their favorites, and the program would notify them when those people connected to the system.

Napster caused controversy even before its official launch. John Fanning, Shawn's uncle and the company's CEO, told the technology news site ZDNet in August 1999 that he meant to run a legal operation, and the company hired the firm of Wilson, Sonsini, Goodrich & Rosati to advise it. This law firm had successfully defended Diamond Multimedia Systems, the makers of the Rio MP3 player, from a lawsuit by RIAA. Napster's primary strategy would be to avoid hosting any materials at all. Instead, it would link computers directly with each other, so that all transmission occurred between the users and their ISPs, not through Napster's own servers. For this reason, Fanning told ZDNet, "There is no copyrighted music that crosses the Napster network. . . . We are about building music communities, not stealing."[20]

Later that year, while the software was still in its beta version, Napster issued a user agreement noting that "copying or distributing unauthorized MP3 files may violate United States and foreign copyright laws. Compliance with the copyright law remains your responsibility."[21] The DMCA required Internet service providers to take action on receipt of a copyright complaint, but P2P was far too new for there to be any clarity regarding even the most basic matters, such as whether Napster or its users counted as ISPs.

Silicon Valley lawyers doubted that copyright owners would sue Napster for making the program. One music industry representative suspected that Napster could be required to "produce system logs that would prove and identify the alleged, so-called piracy."[22] There were also security issues: Was it safe for users to give complete strangers access to their music files? Would intermediary Internet providers, such as universities, allow outsiders to download material from their networks?

A few skeptics thought that these security issues alone were enough to prevent the program from taking off, but the "MP3 matchmaker" (as an MTV reporter dubbed it) was an overnight success.[23] University dormitories, which often had Internet speeds much higher than a standard dial-up connection, became hot spots for Napster usage. By January 2000, Napster had become so popular on college campuses that some universities asked their ISPs to block all traffic coming from Napster. At Hofstra University, for example, Napster usage was so heavy that it slowed down the campus network, and Lanny Udey, the associate dean for information technology, worried that students were misusing university resources for a "commercial service."[24]

FIGURE 5. Napster runs on an iBook computer in 2001, on Mac OS 9. This image was posted by user Njahnke on Wikimedia Commons, and I modified it only to increase its resolution. It is licensed under the Creative Commons Attribution-Share Alike 4.0 International License (https://creativecommons.org/licenses/by-sa/4.0/deed.en).

Users gravitated toward Napster for negligible costs, convenience, and ease of access. A CD's packaging and accompanying materials could motivate a user to purchase a new album, but Napster made the music itself available, free of charge, even before the album's release date. Connecting to a P2P network to search for older music that stores may not be carrying anymore was much easier than visiting a brick-and-mortar retailer only to find that one needed to place a special order. These traits were especially attractive to university students, for whom computers connected to high-speed networks were becoming central sources of entertainment and media acquisition (figure 5).

By May 2000, more than a million users per day were logging into Napster. Its users took pride in the speed with which they could build vast libraries. One freshman at the University of Oregon, sitting in his dorm room, downloaded 275 songs in less than two hours. "That's three days of continuous music," he told a *New York Times* reporter.[25] A sophomore at the University of North Carolina told another *Times* reporter that "with my roommate's CD burner I can make a compilation of whatever I want to play in my car. . . . My brother still buys CDs, but he lives at home with a slow connection."[26] A student at the University of Southern California noted, "What I download is hard to find or does not exist online. . . . If I could find the songs I want online, and if they were offered for sale at a reasonable price, I'd be happy to buy them. The distribution method would hopefully be far easier than hunting stuff down illegally."[27]

Napster allowed students to create large playlists that they could use at parties without having to purchase individual CDs and rip the songs from them individually. The university networks' high speeds allowed students to share songs almost instantaneously, with no need to search for the songs online. According to a student at Yale, "Whenever I'm listening to music, I'm using my MP3 player. . . . Even at parties, people hook up computers to big speakers. It's unfair to have this music out there, yet at the same time it's very convenient. It's a little hypocritical, I have to admit. It's a moral dilemma every time I download a song."[28]

Legal scholars and commentators saw the download craze as part of a broader ethical breakdown in the digital world. Don Tapscott, the author of the best-selling book *Growing Up Digital*, argued that unrestrained P2P downloading was an inevitable next step for a generation of computer users who had grown accustomed to free trials. "They want to try before they buy, because the demo is deeply ingrained in their culture," he told a reporter. "Do they know it's illegal? Yes, maybe. But there's nothing inherent in the Net that transmits good values to kids."[29] Mark Pastin, the president of the Council of Ethical Organizations, insisted that P2P music downloads were evidence of a long-term breakdown of ethical norms of music consumption. Copyright, he said, "has always depended on a bit of an honor code," but now that code "has been shattered."[30]

INFRINGING USES

In 1998, Hilary Rosen and her colleagues at RIAA launched "Soundbyting," a campaign to help university and college administrators discuss music copyright with students. The campaign's materials emphasized that "reproducing and distributing music illegally is akin to stealing."[31] Soundbyting's website contained dozens of bullet points outlining the basics of copyright law, identifying the fines and jail time associated with copyright infringement, and dispelling myths about the relations among music, Internet technologies, and the recording industry. It denied the claim that RIAA was running away from technology and tried to dispel the misguided assumption that posting certain disclaimers along with an upload could release the uploader from infringement liability.[32]

The Soundbyting campaign also involved contacting colleges and universities to notify them of their students' copyright violations. The Recording Industry Association complained to the University of South Carolina at Spartanburg, for instance, that a student was distributing pirated MP3s he kept on his PC. This was a difficult public image situation to navigate. The director of antipiracy for RIAA, Frank Creighton, periodically insisted that they were not threatening to sue anyone but were only requesting that the material be removed and that the university decide what action to take against the student.[33] Gary Kendrick, vice chancellor of information technologies at USC Spartanburg, told *Wired* a somewhat different story: RIAA had "made the university aware of the potential liability and cost of the student's actions."[34] In response, the university installed the Kinnetics Management System—a network management system by Loran Technologies that automatically detected high data usage by individual computers and limited streaming activity. According to Kendrick, this system allowed RIAA to find heavy traffic, identify which traffic belonged to the student, and work with the university "to assure that it would never happen again."[35]

Other universities across the country were also following RIAA's lead in cracking down on music piracy. Carnegie Mellon suspended seventy-one students' Internet access because they used illegal MP3 files, and a student at the University of Oregon, Jeffrey Gerard Levy, was sentenced to two years of probation, frequent urine tests, and limited Internet access under the No Electronic Theft Act—the first MP3 pirating conviction under the act, though he had also uploaded movies and software that one federal agent valued at $70,000.[36] Creighton insisted that RIAA was not targeting college students; it just happened that their team of web specialists, automatic web crawler, and community tips had led them to college campuses.[37]

That November, RIAA announced that it would file a lawsuit against Napster. Lydia Pelliccia, a spokeswoman for the trade group, reported that it had "spent many days sampling the Napster community, and found that virtually all file traffic is unauthorized."[38] The association's general counsel, Cary Sherman, explained that "Napster is about facilitating piracy and trying to build a business on the backs of artists and copyright owners."[39] This development did not surprise anyone at Napster, whose executives and developers insisted that they were free from infringement liability because their servers did not store, or even transmit, the music files that users were downloading. Eileen Richardson, a venture capitalist who served as Napster's CEO from September 1999 to February 2000, would later summarize the company's case by saying, "Just because you are a company that makes crowbars doesn't mean that you're responsible when one is used to break into a house. . . . We are not the MP3 police."[40]

Despite its legal woes, Napster continued to grow, in no small part thanks to a $15 million investment secured by Hummer Winblad Venture Partners. The venture capital firm also named Hank Barry, a former musician and record producer who was also a copyright lawyer and partner at Hummer Winblad, to be Napster's interim CEO. Barry celebrated Napster's achievement in reaching ten million registered users and becoming immensely popular among college students, and although he vowed to "vigorously pursue all pending litigation," he also appeared eager to find a business model that "works for everybody"—including both recording executives and consumers.[41] Having only raised $2 million in funding in 1999, the thirty-three-employee company was finally ready to grow, with Richardson staying as an adviser.

As 2000 opened, Napster's website was telling users that "compliance with the copyright law remains your responsibility." One of its pages, titled, "What You Should Know about Napster and MP3," explained that even though thousands of MP3 files had been "authorized for distribution over the internet by copyright owners," some "may have been created or distributed without copyright owner authorization." Users were therefore advised that neither the existence of the MP3 file nor its listing in the Napster software "indicates whether a particular MP3 has been authorized for copying or distribution." Unauthorized distribution might "violate United States and foreign copyright laws."[42]

For a company under attack for facilitating mass copyright infringement, Napster quickly became extraordinarily interested in its own intellectual property. Changes to the website made a few months after RIAA filed its lawsuit included the addition of a "Terms of Service" page that outlined users' responsibilities toward Napster's IP. In February 2000, for example, the terms included the following section: "All Napster web site design, text, graphics (including button icons and images, but excluding those graphics to which third parties retain copyright ownership), the selection and arrangement thereof, and all Napster software are Copyright © 1999–00 Napster Inc."[43] The terms of service also noted that patents for Napster's software were pending, and it included a partial list of the company's copyrights as well as a section on users' copyrights regarding other people's creations. Users agreed not to do any of the following:

(i) use the Napster service to infringe the intellectual property rights of others in any way;

(ii) use the Napster browser or service, or attempt to penetrate, modify or manipulate the Napster browser or service or any of the hardware or software thereof in order to: invade the privacy of, obtain the identity of, or obtain any personal information about (including but not limited to IP addresses of) any Napster account holder or user, or modify, erase or damage any information contained on the computer of any user connected to the Napster service; or

(iii) reverse engineer any portion of the Napster service or browser.[44]

This section replaced one called "What You Should Know about Napster and MP3." What had been a user advisory became a license agreement intended to shield the company from copyright owners' complaints. Napster was hoping to pass the responsibility for copyright infringement on to the users themselves.

I remember feeling that sense of personal responsibility—or, rather, feeling that there was no way I would get caught. During a trip to the United States, my parents got me a book called *MP3 Underground*.[45] This was a complete guide to music downloads, from the basics of how MP3 files compressed sound to tips on how to manage a collection of thousands of songs. Its appendix was an annotated guide titled "Top 101 Internet Audio Sites." I found it especially helpful that this guide marked sites offering "BYAHO" music (bands you've actually heard of) and links to message boards where I could find peers with tastes similar to mine.

The book also contained a humorous and dire warning: "We tell you everything we can about MP3s and other Internet audio. You don't blame us for system crashes, if none of the programs works the way we said they should, and if the music police smash open your front door with a telephone pole and a SWAT team with their safeties off pinion you against the wall while PC inspectors take your hard drive to be searched for purloined songs."[46]

The authors, father-and-son duo Ron and Michael White, promised on the book's inner cover to be honest and avoid judgment. In exchange, readers had to acknowledge that they were entering "a legal twilight zone" and were responsible for their own lives. The Whites added: "If your immortal soul roasts eternally in hell because you download a Metallica song off the Internet, we just don't care."[47]

The Whites saw the copyright situation as "guerrilla warfare on the Internet." Shawn Fanning had made finding MP3s "as easy as a panty raid," and RIAA was a "musical bully" that was threatening to sue individual college students for infringement.[48] The recording industry's demands that Napster take down individual artists' works were entirely unrealistic because Napster itself did not host any content; users did. A file might be online for only a second, and it is easy to change a file's title and metadata to remove any direct references to the bands involved.

I even remember users creating their own codes to share music under the radar, though today I only remember a code that used leet speech (which replaces vowels with numbers) and used only the first and last three letters of an artist's name. Britney Spears would become BR14RS. As it happens, techniques like this still work. For example, if you visit a video streaming linking site called gaymaletube.com, you can find unauthorized full-length videos by popular OnlyFans performers by playing around with the letters in their usernames: searching for RG, for example, will lead you to a vast collection of videos by popular performer Reno Gold.

The Whites wrote their book in 2000, as Napster's case was moving through the courts, but they weren't particularly worried about it. They wrote: "You know what? It really doesn't matter what the courts decide. It's too late for the record industry to kill off Napster. Like MP3, Napster's a hydra. They might succeed in shutting down Napster's own servers. But that will still leave millions of Napster users out there. . . . Tracking down users or servers would be as effective as stomping on ants to kill a nest. Napster is a system, and a system can live on when the members that make it up are replaced or eliminated."[49]

The reason the Whites were so confident was that everything needed to create new unofficial Napsters was already available. The source code for the program was available online. *CNET,* a major online outlet, reported in January 2000 that a Stanford student named David Weekly had reverse engineered the program and kept a personal website explaining, in detail, how computers exchange information. Weekly's website stated that his findings "will remain here in posterity and to just give you a rough feel of how the protocol works."[50] In a great irony, Napster's attorneys had tried, and failed, to get him to take that information down.[51]

This information had helped Napster become a darling of the open-source movement, the worldwide movement that encourages software creators to issue users a license to copy, modify, and distribute the programs to anyone. This practice is as old as computer programming itself, but it picked up a lot of steam with the formation, in 1998, of a nonprofit called the Open Source Initiative. Several alternatives to Napster were already available, including Wrapster, Scour Exchange, Freenet, Gnutmeg, and Macster, among many others.

MP3 Underground also included a CD-ROM loaded with useful programs, including one called Napigator. Once connected to the Internet, the program would display a list of Napster-like servers, their IP addresses, the number of Napster users connected to each one, and the total number (and sizes) of the music files available from those users. Using this program was easy: if one simply double-clicked on a server's name, Napigator would connect to it. Fanning and RIAA would battle it out in court, but music collectors already had many alternatives.

FROM BETAMAX TO MP3

Few court battles in the history of copyright have received more attention than the music studios' lawsuit against Napster. In a whirlwind of legal maneuvering from 1999 to 2001, Hilary Rosen and her allies coordinated an aggressive attack on Napster that culminated with the Ninth Circuit Court of Appeals' order that Napster stop facilitating the distribution of copyrighted music. Napster, unable to negotiate a licensing deal with the studios, shut down its network and eventually, after a series of corporate acquisitions, became an online music store. I encourage you to watch the documentary *Downloaded* if you'd like to learn more about the case and its cultural and industrial dimensions.[52] Here, I want to focus on a long-standing legal issue that the case reconfigured: the legal restrictions on devices that enable people to commit copyright infringement.

Napster raised several defenses, including safe harbor immunity under the DMCA, its users' noninfringement under fair use, and the record labels' misuse of their copyrights. One of these defenses later became especially important: the "staple article of commerce doctrine," which originated in patent law. Stated simply, this doctrine originally said that selling something that others can use to infringe on a patent can be acceptable if that thing can also be used primarily in ways that do not infringe. It is grounded in section 271 of the Patent Act of 1952, which reads: "Whoever sells a component of a patented machine, manufacture, combination or composition, or a material or apparatus for use in practicing a patented process, constituting a material part of the invention, knowing the same to be especially made or especially adapted for use in an infringement of such patent, and not a staple article or commodity of commerce suitable for substantial noninfringing use, shall be liable as a contributory infringer."[53]

The staple of commerce doctrine became important in copyright law thanks to the Supreme Court's opinion in *Sony v. Universal* (1984), commonly known as the Betamax case. The case concerned home video recording using the Sony Betamax VCRs, a tape-based system that competed with the more popular VHS. Universal and several movie and television studios had sued Sony, claiming that because the system allowed users to record television broadcasts at home, it could be used to infringe on their copyrights. The Supreme Court rejected this argument on the grounds that the Betamax had substantial noninfringing uses such as recording content licensed to free television or whose owners did not mind having it recorded.[54]

Sony v. Universal and several opinions after it raised the issue of identifying the conditions under which users may consume media at times and places different from what the creators intended. In the Betamax case, this problem arose in the form of "time-shifting," the practice of recording a broadcast in order to watch it later. The Court ruled that as long as it was done for private viewing, this was permissible. In 1999, while considering the legality of the Rio MP3 player, the Ninth Circuit had followed the *Sony v. Universal* logic to allow "space-shifting," which it defined as rendering music files portable so the user can listen to them somewhere else. According to the court, the Rio had substantial noninfringing uses other than space-shifting copyrighted works, such as making public domain works portable.[55]

Napster's lawyers hoped that the staple of commerce doctrine would help their defense, but their hopes were misplaced. Unlike the Betamax and Rio systems, Napster was designed to distribute copyrighted material to the public—time- and space-shifting were not its main benefits. According to the Ninth Circuit's opinion: "It is obvious that once a user lists a copy of music he already owns on the Napster system in order to access the music from another location, the song becomes 'available to millions of other individuals,' not just the original CD owner."[56]

Of course, this opinion had an immediate impact on Napster. The court ordered it to monitor and try to control its network to try to prevent future copyright infringement. It also sent the case back to the district court, which would determine how Napster would have to rewrite its software. Napster would not survive this process, especially after it was unable to reach licensing arrangements with music studios.

But the Ninth Circuit's opinion also had an impact on copyright law itself. In a press release for the Electronic Frontier Foundation, Robin D. Gross, a staff attorney for intellectual property, argued that the Ninth Circuit's opinion was "a stark departure" from *Sony v. Universal*.[57] In the earlier case, the Supreme Court had deemed it acceptable to distribute technologies that could enable copyright infringement as long as they had substantial noninfringing uses. But the *Napster* court had added a caveat: this would not hold if the distributor knew of specific infringements and failed to address them.

This reading of the Betamax ruling essentially made it impossible for P2P systems and online sharing networks to operate without some sort of surveillance system to detect and deter copyright infringement. The court's reasoning on this point was clear: "If a computer system operator learns of specific infringing material available on his system and fails to purge such material from the system, the operator knows of and contributes to direct infringement."[58] Napster, the court held, had provided "the site and facilities" for infringement to take place, a practice it justified by citing two outdated cases we discussed in chapter 1: *RTC v. Netcom* (the Scientology case) and *Fonovisa v. Cherry Auction* (the auction case).

The *Napster* opinion gave music, movie, and television studios the ammunition they needed to attack P2P networking in full force. They achieved a major victory on June 27, 2005, when the Supreme Court handed down a unanimous opinion in the case of *MGM v. Grokster*. Like the *Napster* case, this is one of the most highly scrutinized opinions in the history of Internet copyright—and for good reason: *MGM v. Grokster* established, in Justice David Souter's words, that someone who distributes a device "with the object of promoting its use to infringe copyright" is liable for the infringement committed by the device's users.[59]

This was a landmark ruling. The idea that the provider of a technology is liable for users' infringements sounds like a return to the Scientology and *Playboy* cases from the 1990s, when system operators struggled to defend themselves against exactly this accusation. They stressed, repeatedly, that they simply did not have the ability to monitor every file that traversed their servers, and they did not get long-lasting relief until the passage of the DMCA. Now here was a unanimous Supreme Court opinion in 2005 saying that under certain circumstances, system operators could once again be held liable. The difference is that, unlike the networks of the 1990s, several P2P networks were specifically advertised and widely used as tools for massive distribution of copyrighted materials.

Let's take a moment to think about Grokster, one of the defendants in this case, and what it meant to have an "object of promoting" copyright infringement. Launched in 2001 and headquartered in Nevis, West Indies, Grokster was a P2P file-sharing system for distributing any kind of media imaginable: from pictures and music to software and long videos. It promised stunningly fast downloads: a full-length movie, about 400 megabytes, would take between twenty-five minutes (on a T1 connection) and twenty-two hours (on a 56 Kbps modem). Downloading a two-hundred-page novel could range from four seconds to four minutes.[60]

All this came together in "the latest generation of file sharing software."[61] Grokster used SuperNode technology—a term the company used to refer to the high-performance computers in the network that served as its search hubs. Unlike other P2P networks, where any computer in the network could serve this purpose, Grokster reduced search time by directing all search queries to a small subset of computers. It also tracked detailed file information, automatically removed incomplete downloads, and featured an "Auto Resume" function that automatically sought out the same file from multiple sources until the download was complete.[62]

A central component of Grokster's online pitch was its value as a platform for content developers. Its "About Us" section read,

Content developers and owners may now easily broadcast their files through the Grokster Network. You can now publish your work and easily share it with a global audience.

It is easy to publish your work: your family photos, home videos, academic reports, travel journal, diary, recipes, music from your own band—your imagination is the limit. The only thing you have to do is put your files into a folder on your own hard disk![63]

Grokster maintained partnerships with media companies to help users distribute their content through the network. Musicians could partner with a record label and distributor called GigAmerica, while filmmakers could partner up with ReelMind, which offered streaming media sites. Grokster even offered to promote independent content in its website and newsletter— a clear effort to establish noninfringing uses of its technology.[64]

On paper, Grokster had a very strict copyright policy—but one that involved overcoming so many hurdles that enforcing it would be very difficult. The company's terms of service noted, in all capital letters:

WE . . . ASK YOU TO PAY SPECIAL ATTENTION TO AVOID VIOLATING COPY-RIGHT LAWS AND REGULATIONS. AS A CONDITION TO USE THE GROKSTER PRODUCTS AND SERVICES, YOU MUST AGREE THAT YOU WILL NOT USE GROKSTER TO INFRINGE THE INTELLECTUAL PROPERTY OR OTHER RIGHTS OF OTHERS IN ANY WAY. UNAUTHORISED COPYING, DISTRIBUTION, MODI-FICATION, PUBLIC DISPLAY, OR PUBLIC PERFORMANCE OF COPYRIGHTED WORKS IS AN INFRINGEMENT OF THE COPYRIGHT HOLDERS' RIGHTS.[65]

The company also had a "Repeat Copyright Offender Policy," under which system administrators would terminate the accounts of users who violated third parties' intellectual property rights. This policy relied on Grokster receiving "actual knowledge"—that is, a ruling from a judge stating that the user had infringed on someone else's intellectual property rights.

Grokster also had a system that allowed content owners to report repeated copyright infringement by individual users. After receiving a report at abuse@Grokster.com, Grokster would ask the user to file a counter-argument or take down the files. The user would have forty-eight hours to reply, or else their account would be permanently suspended, and the copy-right holder would be able to sue the alleged offender directly. Importantly, since Grokster was not based in the United States, the company and its users were not bound by the provisions of the DMCA. This meant that it had to decide for itself whether individual users were committing copyright infringement under whichever jurisdiction applied to them, and the remedies available to US copyright owners were largely restricted to Grokster's decision on how to handle the case.[66]

Court records suggest that Grokster's staff was very aware that copyrighted files were circulating freely on the network. Exchanges between users and support@grokster.com, the company's general help email address, supported this theory. On February 10, 2002, for example, a user named Tyler Nilson wrote to complain that he couldn't find several games (*Black and White*, *The Sims Hot Date*, and *Tribes 2*). The games were there when he had checked on a dial-up connection, but after he upgraded his home Internet, he couldn't find them anymore.[67]

The support staff wrote that they "have absolutely no control over what users share nor do we know what they are sharing" and advised him to keep trying. Tyler angrily wrote: "Is there anyway to get it from your server de to do that is where the lies are kept or do you have your own download (Black and White) that I can get ALL I WANT IS THE GAME BLACK AND WHITE!!!!!!!!!!!!!!!!!!!! I really don't know how you don't have a way to do this." Support responded: "We do not have a central server for the network. If we did, we would be out of business just like Napster. And why are you complaining? Everything that you get out of our network is free. You could always do the legal thing and go buy the game, we suppose?"[68]

The court records I examined do not suggest that Grokster denied the veracity or accuracy of these exchanges. For what it's worth, I remember *Black and White* and completely understand Tyler's raging desire to play it. I bought my copy. The games I downloaded from Grokster were mainly the ones that needed a console: Super Nintendo, Gameboy, and Sega Genesis games.

The Supreme Court's record for *MGM v. Grokster* includes several other email exchanges, mostly less heated. From their small sample, it appears that most users' questions for support staff were about how to play the media they downloaded and how to burn movies and music onto CDs. Staffers also replied several times to explain that Grokster had no control over the viruses users were downloading onto their computers. One user who was particularly sad about this wrote, "You have a virus (loveletter) under Elton John and Billy Joel titled I guess that's why they call it the blues."[69]

Grokster, Napster, and several other P2P sites perished in 2005, in the aftermath of *MGM Studios v. Grokster,* but the battle against online infringement did not end. Now that the Supreme Court had provided its condemnation of the networks, movie, music, and television studios intensified their efforts to target users. There are plenty of horror stories about how vulnerable users received aggressive threats from these studios. But something else was going on: the studios were willing to target not only individuals sharing music online but the practices and legal protections that allowed them to remain anonymous.

The Recording Industry Association had announced in 2003 that its members were preparing to sue users who shared music on P2P networks. Over the next few years, and especially after the Supreme Court's *Grokster* opinion, these suits became so numerous that the EFF labeled them collectively *RIAA vs. The People.* The advocacy group started keeping close tabs on RIAA, regularly summarizing its findings in a report it updated until 2008.[70]

To sue individuals, RIAA's members had to overcome a problem of privacy: their lawyers could not obtain the names and addresses of P2P network users without requesting the information from users' Internet service providers. The DMCA gave them a special subpoena power that, according to the EFF, RIAA's "lobbyists had slipped into" the bill, section 512(f), which established that "a copyright owner or a person authorized to act on the owner's behalf may request the clerk of any United States district court to issue a subpoena to a service provider for identification of an alleged infringer."[71] This meant, in short, that *accusing* someone of copyright infringement was enough to force an ISP to violate that user's privacy.

According to the EFF, music studios filed 5,460 lawsuits against university students in 2004. The next year, they filed roughly 700 John Doe lawsuits every month, with an unusual peak in April of 1,130. The EFF estimates suggest that by the end of 2005, RIAA had filed at least 11,651 suits and that defendants usually paid between $3,000 and $11,000 to settle. Settling tended to be the cheapest option. Paying for legal representation can quickly become expensive, and at least one court entered a judgment of $6,200 by default against a woman who had ignored a studio's lawsuit.[72]

The studios' complaints against individuals first arrived as very aggressive form letters. One such complaint, obtained by the Electronic Frontier

Foundation, accused the defendant of causing "great and irreparable injury that cannot fully be compensated or measured in money" and stated, ironically, that the studios "have no adequate remedy at law." In addition to demanding money, the complaint asked the court to order the defendant to destroy all copies of the recordings in their possession.[73] This last request shows the real strategy at play: destroying one music file from a computer connected to a P2P network does nothing, because in just one second thousands of other users could download the same file. The suits were meant to make examples of copyright infringers.

Every time I teach the Napster saga to my students, I remind them that the story is important nowadays as a reminder of the limits of the law in curbing our online behaviors. A service may be shut down, but users find a way (to echo *Jurassic Park*). They will find ways to connect with each other and share materials until someone creates a new platform that centralizes their efforts. I remember this so well: Napster's end felt like a blow to my online habits back in the day, but I quickly forgot about the service because there were so many other ways for me to get the music and videos I wanted.

Three years after Napster was shut down, an online entrepreneur who goes by Kim Dotcom founded a Hong Kong company called Megaupload. Megaupload was not a P2P service, but it allowed users to exchange very large files with one another. It had several subsidiary sites. Megaupload had music, but the site I best remember was megavideo.com, where people uploaded full blockbuster movies, porn videos, and episodes of television shows. To watch one of these videos, all you needed was to know where to find the link to it. If you joined the right message boards or (later) Facebook groups, you could find people's playlists and enjoy the show.

Megaupload's saga, like Napster's, was a legal spectacle. In 2012, the US Department of Justice seized the domain names and indicted Kim Dotcom (figure 6). A court in New Zealand (where he was living) allowed him to be extradited. You can learn more about this by watching a fascinating documentary about his life and work titled *Kim Dotcom*.[74] But once again, users found a way. Megaupload was rebranded as mega, and as far as I can tell, mega.io is serving some functions of Megaupload. I noticed this recently by trying to track down pictures and videos by an OnlyFans creator without having to create an account in any porn exchange or using a popular kind of website that scrubs OnlyFans content. A creator from Chile was charging

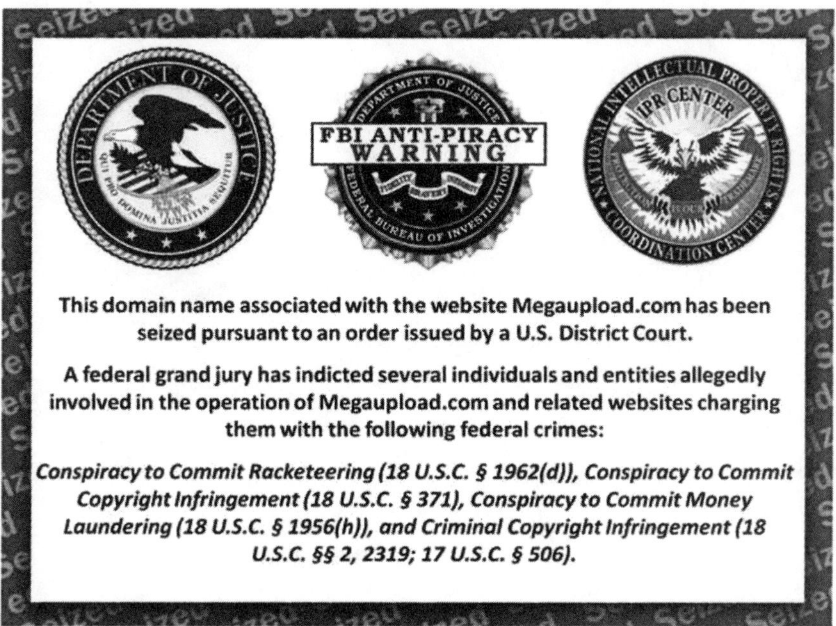

FIGURE 6. An FBI warning that users encountered when they visited Megaupload.com after
the indictment. This image is in the public domain.

$12.99 a month for sexual content and placed stern warnings on his
OnlyFans page about copyright infringement. It took me a whole three
minutes on Facebook and Reddit to find a mega.io link to a folder with all
his content.

The path I took to this content was the scenic route because an even
easier way of getting any content is to log into P2P networks. Some of these
networks, such as BitTorrent, can have any kind of content you could dream
of. There are also specialized networks for pornography. Sure enough, my
OnlyFans creator's content was found in a few of them. I certainly advise
you *not* to use these networks, since Internet service providers can track
what you upload and download. However, several folks I know use them
thanks to combination technologies that mask their IP addresses (which can
help identify your location). Don't do it yourself—but know that Napster's
death was just the beginning.

On January 18, 2012, a remarkable incident took place: Google, Wikipedia, Reddit, and other Internet giants blacked out their content. Scores of other sites did the same, draping their logos with black rectangles and blocking at least some pages. They meant to convey the message that the Internet was in danger. The US Congress was considering two bills introduced in 2011 that would have significantly bolstered copyright law's reach on the Internet. From New York to Moscow, advocacy groups, law and policy professionals, and technology companies took their complaints against the legislation to the streets and the digital world.

I was working toward my doctorate and was just starting to prepare for something called qualifying exams. I had four exams to prepare for, each one encompassing up to sixty books. I was just becoming interested in intellectual property as a field of research, and when I read that a copyright-related Internet blackout was coming, I forgot about my exams and went down an Internet rabbit hole of censored content—in the hope that one day I would get to write the book you are reading right now.

My daylong excursion started at www.google.com, where I was greeted by a black rectangle blocking all but the bottom curve of the second g in the company's name. Under the search bar was a message: "Tell Congress: Please don't censor the web!" Clicking on the link led to a simple page warning that the two bills "would censor the Internet and slow economic growth in the U.S." This page also had a sort of petition on the right-hand side, which folks could sign to oppose these bills. (You can visit this page via the note cited at the end of this sentence, and I invite you to go on your own excursion of blocked content by typing in the URL for your favorite website at the search bar with the text https://www.google.com/landing/takeaction/.)[1]

The two bills were the Stop Online Piracy Act (SOPA) and the PROTECT IP Act (PIPA). Both were backed by large organizations of copyright owners such as Hollywood studios and music labels: SOPA was introduced in the Senate by Lamar Smith (R-Texas), PIPA in the House by Patrick Leahy (D-Vermont). They were remarkably similar in intent—both were efforts to end the circulation of copyright-infringing material and counterfeit goods—and were backed by such powerful groups as the Motion Picture Association of America, the Recording Industry Association, and the US Chamber of Commerce. The bills were so important to these groups that in just the closing quarter of 2011, corporate supporters spent $14.2 million lobbying for them.[2]

The story of SOPA and PIPA has been told many times. My favorite account comes from a law professor, Edward Lee, who moved very quickly while events were unfolding and in 2013 published a book titled *The Fight for the Future*.[3] With the speed and succinctness of a journalist, Lee showed how SOPA and PIPA were the American embodiments of an antipiracy initiative, Anti-Counterfeiting Trade Agreement, or ACTA, that the United States launched among its trading partners in 2007. The public outcry against the bills, he implies, was the local manifestation of a broader international outcry against this initiative. For this reason, instead of walking you through the rise and fall of SOPA and PIPA, I prefer to reflect on how the bills and their failure fit within the larger story of online copyright.

SOPA and PIPA were bold bills. They empowered judges to prohibit advertisers and payment processors from doing business with websites accused of copyright infringement, bar search engines from linking to those sites, and even block access to them entirely. They also imposed harsh criminal penalties from streaming unlicensed content, including up to five years in prison. The PROTECT IP Act targeted "rogue websites dedicated to the sale of infringing or counterfeit goods," including illegal copies or technology meant to bypass digital rights management systems. It empowered the US attorney general to pursue court orders against foreign websites, which would not normally fall within the Justice Department's jurisdiction.

To grasp the potential impact of SOPA and PIPA, it is helpful to revisit a foundational metaphor for the Internet. You may remember how the court in *RTC v. Netcom* likened the Internet to a swap meet while weighing a copyright infringement case brought by the Church of Scientology. Years before the DMCA established protections for Internet service providers, the swap meet metaphor played a crucial role on debates over online copyright infringement liability. It allowed the court to reject the argument that a bulletin board system was directly and vicariously liable for its users' copyright infringement, for the same reason that the owners of a space where a swap meet takes place are not liable if some of the swappers exchange stolen material.

Imagine a town with a bustling swap meet, where artists of all kinds come to display and exchange their creations. Photographers, writers, musicians, and others set up booths to distribute and sell their creations. These artists often copy each other's work for sale in their own booths, and they don't always ask permission. Both SOPA and PIPA were grounded on the assumption that the swap meet had a growing problem with counterfeiters, many of whom had set up booths exclusively to sell unauthorized reproductions of other creators' original art. The acts would have granted the town's authorities the power to shut down all booths where counterfeit art was being distributed without permission, prohibit these booths from making any financial transactions, and investigate any sellers suspected of counterfeiting activities.

If the town's authorities exercised their full power, not only would all counterfeit booths vanish, but so would the booths of any artists who displayed even one unauthorized work by another artist. A booth featuring the most exquisite portraits could be shut down because its promotional materials used an unauthorized landscape as a backdrop. A booth for swapping music could suddenly be under investigation because one of its songs bore a suspicious resemblance to another. And no booth would be able to offer a preview of the swap meet without risking an investigation.

Both SOPA and PIPA would have strengthened and expanded the enforcement mechanisms for online copyright to the point that they could easily lead to preemptive censorship. The bills allowed copyright owners to get court orders requiring websites and platforms to block access to sites or domains *accused* of hosting or linking to infringing content. Websites and platforms would have had to censor content based on allegations, without any due process. The most troubling implications, however, concerned user-generated content. The fear of being accused of copyright infringement would have chilling effects, both on users' ability to express themselves online and on platforms' willingness to display or distribute original content, for fear that they might share liability for their users' copyright infringements.

The bills would have also granted broad immunity to Internet service providers who blocked users voluntarily. Service providers would need only to act "in good faith" and "on credible evidence," and they would not need any court oversight. This could have had a catastrophic impact on competition among Internet companies. For example, a studio could block an entire video-sharing platform by claiming it had credible evidence (such as a user activity log) that the platform was being used for infringement. It would then be up to the platform to prove that it was innocent or that the accuser had acted in bad faith. And while that question was being adjudicated, the site would be taken down.

It is not surprising that Google and its lobbyists led the charge in trying to defeat SOPA and PIPA. The bills were an important threat to search engines and online platforms, particularly Google, which was already facing an antitrust investigation and several battles over privacy. In response, Google became a lobbying powerhouse, spending approximately $9.7 million on lobbying during 2011 (twice what it had spent in 2010), with a third of it spent in the final quarter of the year. This amount even surpassed Microsoft's lobbying expenditures by $2.4 million.[4]

Eric Schmidt, Google's executive chairman, issued a statement condemning SOPA and PIPA because they would "criminalize linking and the fundamental structure of the Internet Itself."[5] The DC news outlet *The Hill* quoted him as saying, "It's not a good thing. I understand the goal of what SOPA and PIPA are trying to do. . . . Their goal is reasonable, their mechanism is terrible. They should not criminalize the intermediaries. They should go after the people that are violating the law."[6] Google founder Sergey Brin told *Politico,* "At various times notable Google websites have been blocked in China, Iran, Libya (prior to their revolution), Tunisia (also prior to revolution) and others. . . . Thus imagine my astonishment when the newest threat to free speech has come from none other but the United States"[7]

These hyperbolic, all-or-nothing condemnations reflected Google's clear, high-stakes financial interest in defeating SOPA and PIPA. But their tone shows how public discourse about the bills could devolve into the hyperpolarized posturing that is all too familiar in US politics.

An Internet activist group called Demand Progress, for example, posted an announcement on its website that read, "URGENT ACTION NEEDED: New Internet Blacklist Bill Could Shut Down Twitter and YouTube!" Without once mentioning the word "copyright," the site warned that the bills "could effectively destroy Youtube [*sic*], Twitter, Facebook" and all other sites that featured user-generated content.[8] A blog post at the US Chamber of Commerce by the prominent IP attorney Steve Tepp responded with a warning: "Don't Be Fooled by Anti-IP's Scare Tactics." "We will be highlighting the most extreme and absurd claims made by the anti-IP crowd," Tepp wrote, "and debunking them with *gasp* . . . the facts."[9]

Both SOPA and PIPA had the potential to cause widespread censorship in the name of copyright. This was especially clear at the Wikimedia Foundation (figure 7), which in October 2011 took a bold public stance against an Italian bill proposed by the administration of Prime Minister Silvio Berlusconi.[10] According to the foundation, the bill would require "all websites to publish, within 48 hours of the request and without any comment, a correction of any content that the *applicant* deems detrimental to his/her image." In response, the foundation blocked all its Italian-edition content and issued a public condemnation of the bill, signed by "Gli utenti di Wikipedia" (Wikipedia's users).[11]

On December 10, 2011, after the Italian bill failed, Wikipedia cofounder Jimmy Wales explained in a blog post that "a much worse law" (SOPA) was on the congressional fast track.[12] He proposed a straw poll to gauge user support on blacking out Wikipedia once more. It was a brilliant move: there was little doubt that Wikipedia users would side with him, but asking folks to vote and comment allowed him to spread the word and generate a record with thousands of entries. There was overwhelming support for blocking access to the site, and many of those who opposed this action did so not in opposition to SOPA or PIPA but because they believed that the matter did not concern Wikipedia.

A few days later, an anonymous user who signed as "J. S." created a whitehouse.gov petition to illustrate just how damaging SOPA and PIPA could be. The petition was very short and is worth quoting in full:

FIGURE 7. This is the screen that greeted Wikipedia visitors on January 18, 2012. It was uploaded to Wikimedia Commons by user Tomwsulcer, and I modified it only to increase its resolution. It is licensed under the Creative Commons Attribution-Share Alike 3.0 Unported License (https://creativecommons.org/licenses/by-sa/3.0/deed.en).

WE PETITION THE OBAMA ADMINISTRATION TO:

VETO the SOPA bill and any other future bills that threaten to diminish the free flow of information

"the more freely information flows; the stronger that society becomes"

President Obama http://tiny.cc/rh5b1

By allowing free conversation it is so easy to drop a link

http://i.imgur.com/TD4Kq.jpg[13]

It would be ridiculous for an ISP to block the entire whitehouse.gov domain on court order because a single user posted a link. It is difficult for any web administrator to know which links to copyrighted material are done with permission. This will kill the free flow of information and conversation on the internet.

SOPA is too blunt. Please veto.[14]

The petition's statement quoting President Barack Obama comes from a speech that the president made to a group of students in China earlier in 2011: "I can tell you that in the United States, the fact that we have free Internet—or unrestricted Internet access—is a source of strength, and I think should be encouraged. . . . The more freely information flows, the stronger the society becomes."[15]

As the petition noted, the Obama administration had made openness a top priority.[16] Earlier that year, White House representatives had gathered with representatives of thirty-four countries and Internet pioneers such as Tim Berners-Lee and Vincent Cerf to draft a set of Internet policymaking principles. These principles included promoting and protecting "the global free flow of information" and limiting "Internet intermediary liability."[17] The White House was quite blunt in saying that efforts to obstruct international content flow "risk balkanizing the Internet" and had "high costs in both economic opportunity and the realization of human rights."[18]

All this meant that opposing SOPA and PIPA aligned with White House priorities, but J. S.'s point in the petition was to illustrate that if SOPA and PIPA passed, even the petition would be shut down. The first of its two links is a temporarily shortened URL that now leads to a website on how to kill deer and display their heads on a wall. But the second directed users to the image in figure 8. It was created in 2011 by Mike Fahmie and Jeremy Vinar, two cartoonists who posted their work at www.virtualshackles.com. It depicts a man resembling Abraham Lincoln sitting inside a prison cell alongside the Reddit mascot. Both are wearing orange jumpsuits, and a poster of a US flag on the wall behind them is overcast by the cell bars in the foreground.

This image, like many others I remember circulating online at the time, includes text stating that reposting it would be a crime. The text on the top left says that SOPA and PIPA would have allowed up to five years in prison and a hefty fine for each infringement. The text on the bottom right critiques Congress's high defense expenditures and threatens to hold the viewer in jail indefinitely.

Outcry against SOPA and PIPA reached a new high in January 2012 when the White House answered J. S.'s petition and a similar one targeted at another bill (figure 9).[19] Three officials responded: Victoria Espinel (intellectual property enforcement coordinator at the Office of Management and Budget), Aneesh Chopra (US chief technology officer), and Howard Schmidt (special assistant to the president and cybersecurity coordinator for National

FIGURE 8. Mike Fahmie and Jeremy Vinar made one of my favorite images in protest of SOPA and PIPA. They granted me permission to include the image in the book and requested the following credit: Mike Fahmie and Jeremy Vinar.

Security staff). The response, which the EFF called "potentially game changing," took the censorship claims seriously.[20] "Any effort to combat online piracy," it stated, "must guard against the risk of online censorship of lawful activity and must not inhibit innovation by our dynamic businesses large and small." It also acknowledged that court-ordered blockages of websites would disrupt "the underlying architecture of the Internet."[21]

This suggested that Obama would veto SOPA and PIPA if Congress managed to pass them. But the fight continued: the Senate was scheduled to bring PIPA to the floor, and the EFF expected that SOPA proponents in the House would "try to revive the legislation—unless they get the message that

FIGURE 9. A woman sits outside the offices of Senators Chuck Schumer and Kirsten Gillibrand in New York City on January 18, 2012, to protest SOPA and PIPA. This photograph was uploaded to Wikimedia Commons by user RoySmith, and I modified it only to increase its resolution. It is licensed under the Creative Commons Attribution-Share Alike 4.0 International License (https://creativecommons.org/licenses/by-sa/4.0/deed.en).

these initiatives must stop, now."[22] And they were right: the Motion Picture Association of America tweeted, "Look forward to @whitehouse playing a constructive role in moving forward on #sopa & #pipa," and media mogul Rupert Murdoch tweeted, "So Obama has thrown in his lot with Silicon Valley paymasters who threaten all software creators with piracy, plain thievery."[23]

The Great Internet Blackout happened right after the White House's response, but it had been in the works for several months. Back in November 2011, a personal blog company called Tumblr added a tab called "censored" to its dashboard that directed users to the bare-bones content they'd see if the bills had passed. The company hosted a meeting in its Manhattan offices for employees and outside executives to discuss the problems with the bills. The discussion eventually moved to Reddit, where users identified GoDaddy (a domain name seller) as a supporter of the bills and coordinated a boycott. Twitter picked up on the story, and the number of tweets mentioning SOPA

FIGURE 10. This illustration by Chris Piascik sums up the lighthearted way I introduce my students to what SOPA and PIPA could have done to the web. Chris Piascik granted me permission to include this image in the book and requested the following credit line: Chris Piascik.

rose from 200,000 two days before the protest to 3.9 million on the day of the blackout. A nonprofit organization called Fight for the Future helped to organize protests. As Internet scholar Clay Shirky noted at the time, through these actions "a disorganized group of people online became a coordinated group of people taking action."[24]

There was plenty of humor. While companies and trade associations made grand statements and commenters waged online firestorms, illustrators gave us countless opportunities to laugh. New England cartoonist Chris Piascik, for example, reminded us that the future of funny cat memes was at stake (figure 10).[25] But the point was much more serious than that: if bills like SOPA and PIPA became laws, user-generated content could cease to flow.

This brings us back to the blacked-out Google page. The day after the protests, Google announced that it had collected more than 7 million signatures from the United States, and Wikipedia reported that more than 162 million people had seen its own blacked-out site. *Bloomberg* reported that thirteen lawmakers who had cosponsored SOPA had withdrawn their support, eight in the Senate and five in the House.[26] Republicans and Democrats were both backing out.

The SOPA and PIPA bills were ultimately defeated, but their story shows that users, copyrights, and the Internet are woven so tightly that they have become inseparable. The fate of the Internet was at stake when these bills were in Congress—not only in the United States but around the world, as users and content makers in other countries organized to push back similar bills.

Cracking the Code

IN FEBRUARY 1999, REPRESENTATIVES FROM more than two hundred companies in the worldwide music industry gathered in Los Angeles, California, to launch the Secure Digital Music Initiative—SDMI for short.[1] The goal was to create "an open architecture and specification for digital music security" that would allow music files to be transported from one device to another if, and only if, the user was authorized to do so. The fact sheet for SMDI stated that the initiative "is based on the core principles that copyrights should be respected and that those who wish to do so should be able to use unprotected formats."[2]

No one fit the bill for SDMI's executive director better than an Italian electrical engineer named Leonardo Chiariglione. Dubbed "Father of the MP3" by *Time*, he is best known for cofounding the Motion Picture Experts Group (MPEG), a technical consortium that created the file formats and compression that make large multimedia files portable, including MP3's. Chiariglione felt a distinctly Catholic and capitalist calling to the music world. Reflecting on his career, he explains: "My Christian Catholic education made and still makes me think that everybody should have a mission that extends beyond their personal interests. In my early professional years, when digital media technologies were still maturing, I saw my mission in *the development of interoperable digital media technologies for* society *to enjoy and industry to exploit.* . . . I needed an organisation that would *create digital* *media standards for **consumers** to seamlessly communicate and **industry** [to] operate in a global market of interoperable products, services and applications."*[3]

Despite its celebrated launch, SDMI had a short, bumpy history.[4] In the summer of 1999, it issued a thirty-five-page document explaining that equipment manufacturers would go through two phases in their long-term goal to produce and sell SDMI-compliant devices. Devices produced during the first phase would play music in all formats, regardless of whether those formats enabled any kind of copy protection. The second phase would require the production of devices with screening technology to identify and disable pirated files. Consumers would have to voluntarily upgrade to second-phase devices—a move the music industry would incentivize by releasing all future music in a format that only these second-generation devices could play.[5]

As Tarleton Gillespie writes in *Wired Shut,* the goal of this rollout was the planned obsolescence of devices that couldn't identify and disable pirated content. The second phase would also involve a new generation of intercompatible music players that encompassed everything from CD and MP3 players to personal computers. According to SDMI, this would generate a market of "devices that respect music industry copyrights, systems that accept protected and unprotected music, and flexible systems that allow everything consumers have today."[6] For all of this to work, SDMI would need to test its piracy detection systems' ability to stand up against hackers.

In the fall of 2000, Chiariglione posted an online challenge to anyone interested in testing their hacking chops: beat SDMI's most advanced antipiracy technologies in a month and win $10,000. The initiative was moving to content containing either an "inaudible, robust watermark" or some "other technology that is designed to prevent the unauthorized copying, sharing, and use of digital music." It was testing these technologies to make sure they were "inaudible, robust, and run efficiently on various platforms, including PCs," but it thought users should also test these technologies. "So here's the invitation," Chiariglione wrote. "Attack the proposed technologies. Crack them."[7]

The goal was to test whether the nearly imperceptible sounds that publishers would embed into their audio files could be identified and removed. It worked like this: SDMI made available six technologies, each designed to mark ownership of audio files. Most of them were audio watermark generators—programs that introduced a very faint sound into an audio track. These sounds were meant to be unobtrusive, readable, and indelible: nearly inaudible to the human ear, detectable by the appropriate program, and impossible to delete without damaging the audio quality of the file.

Hackers, computer scientists, and other outsiders were asked to try to alter any of these qualities in an audio sample's watermark. The best possible outcome would be to remove the watermark altogether without altering the rest of the audio file. This achievement qualified as "defeating" the technology and made researchers eligible for the grand prize. But since the technologies created sounds the human ear could not perceive, contestants' success would be determined by a program ominously called the "Oracle." Contestants would upload their efforts to the Oracle and wait for the program to decide whether they were eligible for the prize.[8]

Beating the copyright protection technologies thus boiled down to beating the Oracle. This automated judge was a classic example of what researchers call a "black box." It was a technology known only for inputs and outputs, not for how or why it works. Contestants did not have direct access to its computer code, so they could not figure out what tests it ran on the samples. The only thing they could do was to start uploading samples to the Oracle to see what it could and couldn't detect. If they did that enough times, they could begin to refine their approach to focus only on the things they knew the Oracle cared about.

This invitation was irresistible to Edward Felten, a computer science professor at Princeton and the director of the university's Secure Internet Programming Laboratory. By all accounts Felten was a rising star—winner of the National Science Foundation's National Young Investigator award, an Alfred P. Sloan Foundation Fellowship, and numerous grants and gifts from IBM, Microsoft, and Merrill Lynch. He had also assisted the FBI's investigation of the Melissa virus (which took over people's Microsoft Word documents) and had testified in the Department of Justice's antitrust case against Microsoft.[9]

Felten quickly assembled an all-star team to tackle the challenge. It included Drew Dean from Xerox's Palo Alto Research Center, Professors Bede Liu (Princeton) and Dan Wallach (Rice), and five graduate students from Rice and Princeton. Before they could do anything, though, they needed to handle some legal matters. First, they had to sign a click-through agreement to access the materials for the challenge. This agreement specified that they would have access to SDMI's samples for only three weeks. It also required them to choose one of two options: either retain the right to publish and thereby forego the $10,000 prize, or sign an additional agreement not to disclose their research and thereby collect the cash prize.

Here is what the agreement said: "In order to receive compensation, you will be required to enter into a separate agreement, by which you will assign your rights in such intellectual property. The agreement will provide that (1) you will not be permitted to disclose any information about the details of the attack to any other party, (2) you represent and warrant that the idea for the attack is yours alone and that the attack was not devised by someone else, and (3) you authorize us to disclose that you submitted a successful challenge. If you are a minor, it will be necessary for you and your parent or guardian to sign this document, and any compensation will be paid to your parent or guardian."[10]

No one in Felten's team signed this additional agreement because they planned to publish their results. But they all accepted the click-through agreement and had every reason to believe that they would be allowed to publish their results if they simply declined the cash prize. The click-through agreement specified, "You may, of course, elect not to receive compensation, in which event you will not be required to sign a separate document or assign any of your intellectual property rights, although you are still encouraged to submit details of your attack."[11]

Since they were forfeiting the prize, the team focused on testing the Oracle to evaluate the robustness of SDMI's technologies. The technical challenge was clear (removing watermarks from SDMI's audio clips), but the parameters for success were not. The initiative issued no information about the Oracle's algorithm or evaluation standards, so Felten's team used trial and error to determine what the Oracle would deem a good effort. They submitted a series of audio clips as "experiments to characterize the oracle's behavior" that would help them understand how the watermarking technology worked.[12]

These experiments gave them some useful clues. For example, the Oracle rejected audio clips with a detectable watermark even if the clips had perfect audio quality. It also rejected clips with no watermark and lower audio quality. The team learned that the Oracle would reject a sample if it had even a small trace of the original watermark but that a small amount of reduction in audio quality was acceptable.[13] This suggested that SDMI's technologies were not very robust, and the team defeated the system with ease.

Felten would later recall that the hardest part of all of this would be writing a paper. Many of the contributors had never worked together, and they disagreed on how deeply the paper should state the group's general skepticism about watermarking technologies.[14] But SDMI had given them an even bigger and very unusual challenge by asking them to sign the click-through agreement. This document prohibited further research using SDMI's clips after the challenge ended—at which point the group would also lose access to the Oracle. The initiative offered them an opportunity to participate in a follow-up study, but they declined because they would not have permission to use the data for their own research. This meant that they would not be able to perform any follow-up studies or fill in any gaps they found while writing the paper.[15]

Felten and his colleagues were experiencing the complicated relationships that scientists can have with the law. There's a lot of research on how law, science, and technology affect one another. Sometimes new technologies and scientific developments change how the law is applied. Think about DNA fingerprinting: until this technology arrived in the late twentieth century, it was very difficult to test a bodily substance found at a crime scene to determine whose body it came from.

Felten's case brings up a different relationship: legal circumstances can *hinder* scientific research by blocking access to the resources needed to continue a line of inquiry, restricting the content that can be disclosed in scientific publications, and making it difficult to participate at scientific conferences. Before we move on, however, I should tell you about the broader changes in copyright law that were happening around that time.

LEGAL CODE / COMPUTER CODE

In the fall of 2000, while Felten's team was working on their paper, the District Court for the Southern District of New York published an opinion called *Universal City Studios v. Reimerdes*. This case concerned an encryption system for DVDs called content scramble system (CSS). The DVD Copy Control Association (the film industry's counterpart to SDMI) had developed CSS as a way of restricting the unauthorized global flow of DVDs.[16] Rather than preventing copying, it restricted which DVD players could play a given disk. You may have run into this type of system if you have purchased media in several countries: a DVD or Blu-ray purchased in Europe or China would probably not work in the United States unless you used a program such as VLC Media Player, which is used to bypass zone restrictions.

About a year before the court's opinion, three programmers had developed a program called DeCSS that could bypass this system. Within weeks after its release, it had circulated worldwide through online magazines, personal websites, and message boards. Anyone using a popular open-source operating system such as Linux could watch any DVD regardless of its zone encryption. Movie studios sued the programmers and several parties for violating the DCMA's anticircumvention provisions. The defendants included the magazine *2600: The Hacker Quarterly*, which had shared the code; its publisher, Eric Corley; Shawn Reimerdes of dvd-copy.com, who had posted the code on his site; and Roman Kazan, the owner of an online server from which DeCSS could be downloaded.

Ultimately, the district court sided with the studios. It affirmed that the DMCA prohibits the distribution of software that allows users to circumvent copyright protection, and it rejected Corley's argument that the DMCA itself was an unconstitutional violation of his First Amendment rights. This has been, for many years, a classic defense against accusations of online copyright infringement. It is grounded on the notion that creative works, including computer programs, are protected speech under the First Amendment and that the DMCA illegally prohibits their dissemination. The judge did not disagree with the idea that computer code is a form of speech, but he ruled that it was constitutionally acceptable to restrict its circulation because the goal was to restrict what the code *does,* not what it *says.*

Felten, who was well versed in law and policy, was careful not to expose his team and their institutions to legal liability. He insisted that their paper on the SDMI challenge not include "any software code, pseudocode, or code-like descriptions of algorithms" even though the team firmly believed that doing so would improve the paper's quality.[17] But they all agreed that the click-through agreement gave them the right to discuss the research they had done and disseminate their conclusions about SDMI's technologies.

The team's paper was excellent. They submitted it to the Fourth International Information Hiding Workshop and received very positive peer reviews. One reviewer described it as a "tour de force."[18] The scientists were encouraged to hear that the legal restrictions they were navigating had not led them to write a subpar paper with major holes. They quickly revised the paper in accordance with the reviewers' suggestions and submitted it for inclusion in the workshop's compendium. They also agreed to present the paper at the workshop's meeting in April 2001.

Signs of legal trouble began in November 2000. Felten was in Greece at a computer security conference when he received an email from Joseph Winograd, the executive vice president and chief technologist at the Verance Corporation in San Diego, California. Verance had developed one of the watermarking technologies in the SDMI challenge, and Winograd wanted to know what Felten's team had discovered about it. Felten agreed to speak with Winograd on the phone after he returned from the conference and to send him a prepublication copy of the paper. This was a courtesy he offered whenever he published work on commercial technologies.

In April 2001, Winograd sent an email to Felten saying he was "most concerned" with the paper's contents. He requested another call to discuss the paper and added that he had taken the "precautionary step of alerting the SDMI Foundation" and giving them a "brief general description" of the paper.[19] Three days later, Felten received a letter from Matthew Oppenheim, the vice president for legal affairs at the Recording Industry Association of America and the SDMI Foundation's secretary. The letter threatened legal action against the authors and their universities if the paper was published. Disclosing any information that would make it easier to remove Verance's watermark, or any other mark used in the challenge, "would seriously jeopardize the technology and the content it protects."[20]

On RIAA letterhead, Oppenheim wrote, "Unfortunately, the disclosure that you are contemplating could result in significantly broader consequences and could directly lead to the illegal distribution of copyrighted material. Such disclosure is not authorized in the Agreement, would constitute a violation of the Agreement and would subject your research team to enforcement actions under the DMCA and possibly other federal laws." Publishing the paper would amount to "facilitating and encouraging the attack of copyrighted content outside the limited boundaries of the Public Challenge."[21]

With the Information Hiding Workshop just a few weeks away, Felten had several conversations with lawyers for the researchers' institutions, Verance, and RIAA. On April 26, Winograd sent Felten twenty-five requests for revision, including deletion of entire sections of the text and full diagrams.[22] Felten later testified that following Winograd's recommendations "would have severely gutted the paper by removing virtually all of its detailed technical content." He stated that thought such a paper would be rejected by any reputable journal and would receive "scathing reviews."[23]

The workshop's program committee grew nervous. According to Felten, just a week before their presentation, Ira Moskowitz (the program chair) emailed him to explain that the paper would be removed from the program unless all the parties involved certified that it would be legal to publish it. Moskowitz called Felten late at night to explain that there had been great pressure on him to make that decision.[24]

The full program committee eventually reversed Moskowitz's decision, but Felten and the team agreed that it was best not to present their paper at the workshop. Instead, at their scheduled time, Felten delivered a statement describing RIAA's, SDMI's, and Verance's threats as a danger to free speech and scientific debate. "We will continue to fight for these values," he told his audience, "and for the right to publish our paper."[25]

USENIX

Around this time, legal scholar Lawrence Lessig published a scathing critique of corporate uses of intellectual property titled *The Future of Ideas*. He argued that powerful organizations were aggressively enforcing their intellectual property rights, threatening individual rights and social well-being, and restricting our ability to generate new ideas. He concluded his discussion of DeCSS, for instance, by describing the lesson Hollywood was sending: "Any system that threatens its control will be threatened with an army of Hollywood lawyers."[26]

In hindsight, Lessig's stern warning proved to be more apocalyptic than reality. But Felten's story suggests that there was good reason to be worried at the time. A renowned scientist and his colleagues had responded to a research challenge and followed all the legal procedures required for their participation. They had prepared a top-shelf paper, but lawyers for the music industry and its technological allies were pressuring globally prominent scientific venues that considered hosting the research. This was not the music industry controlling the circulation of music files: it was seeking to control new knowledge.

While searching for a new publication venue, Felten's team submitted their paper to a symposium organized by USENIX, a nonprofit organization founded in 1975 and also known as the Advanced Computing Systems

Association. In addition to its annual conferences on system administration, USENIX hosted specialized events on topics including computer security, e-commerce, operating systems, and object-oriented programming. These events were very selective venues that had already published papers designed to attack emerging or commercially viable security systems.[27]

This was as open a venue as Felten's team could have aimed for. All USENIX papers eventually became publicly available. The organization maintained an extensive public online database of all the papers presented at its conferences. From 1993 to 1998, it had published 848 papers online—an average of 141 per year. By 2001, the site included 1,406 papers.[28] All had undergone rigorous peer review, and many included large portions of code.

Felten's submission arrived a month after the symposium's deadline, but the program committee accepted it to the conference. Executive Director Ellie Young and her committee ranked the paper highly, but she would later note that the threats against the Information Hiding Workshop "were quite frightening."[29] She consulted a lawyer, who warned her that USENIX might be sued if it presented or published the paper, especially because lawyers could now cite the *Universal City* court case.

Young was in a difficult position. On the one hand, she and her colleagues worried that Verance and its allies would target them next. This would be very dangerous because USENIX's conference-centered business model would allow Verance to argue that the nonprofit derived a commercial advantage from violating the company's copyrights.[30] On the other hand, not accepting the paper would mean letting commercial interests interfere in the peer review process. It would amount to censoring research to meet a corporation's demands.

Young opposed this censorship, but she needed to keep USENIX free of liability. This is how she described her team's decision in a legal declaration: "Although the Felten paper has been accepted for the August conference, USENIX will remove the paper from the conference agenda and proceedings if the fear of liability is not definitively eliminated. The bound volume of proceedings will be finished on or about August 10. We will cut the paper out of the bound volumes by hand if we have to."[31]

But Young also planned to fight back. One June 6, 2001, the researchers and USENIX filed a complaint against RIAA, Verance, SDMI, and other developers of SDMI technologies.[32] They were represented by several lawyers (including three from the Electronic Frontier Foundation) who asked the District Court of New Jersey to issue a declaratory judgment stating that "publication of the paper is lawful."[33]

They sought other declarations as well. One was that the researchers had not committed a DMCA violation by submitting the paper for consideration by the conferences and that publishing their work would not be a violation either. Another was that these actions did not constitute violations of the click-through agreement. This was important to everyone involved, but especially to Min Wu, a graduate student on Felten's team who had incorporated some of the research into her dissertation.

The complaint also emphasized free speech. It claimed that the DMCA had chilled the researchers' free speech rights by generating fears of civil and criminal liability and enabling corporations to prevent the publication of the researchers' work. The DMCA, the complaint declared, "wreaks havoc in the marketplace of ideas, not only the right to speak, but the right to receive information—the right to learn."[34]

This is the kind of argument that could be expected from the Electronic Frontier Foundation at this point. The complaint said that "by imposing civil and criminal liability for publishing speech (including computer code) about technologies of access and copy control measures and copyright management information systems, the challenged DMCA provisions impermissibly restrict freedom of speech and of the press, academic freedom and other rights secured by the First Amendment to the United States Constitution." The DMCA would "harm science" unless "full and open access to research in areas potentially covered by the DMCA" could be ensured.[35]

An amended complaint filed in June 2001 argued that the DMCA "exceeds Congress's enumerated powers," since neither the DMCA itself nor its legislative history specified which power Congress was exercising when it enacted the anticircumvention provisions. The law challenged researchers' ability to "make lawful and constitutionally protected uses of technology" because it effectively prohibited "the distribution of technology without regard for originality, duration of copyright, or infringement of copyright in the underlying, technologically-protected work."[36]

Represented by Karen A. Confoy and David E. Kendall, RIAA aimed to show that there was simply no reason to sue them. Confoy and Kendall claimed that Felten and the EFF's "true agenda" was to "obtain favorable press attention and to secure an advisory opinion on the constitutionality of the Digital Millennium Copyright Act." They accused the scientists of filing a complaint "full of vaporous imagining and chimerical fears" and asked that the court dismiss it because there was no "genuine conflict."[37]

The lawyers also insisted that RIAA did not object to having Felten's team present and publish their paper and that the scientists had no reason to worry about any future papers they might write. There was no need for the court to consider "speculative claims based on hypothetical papers, not yet written, to which no one has ever objected."[38] Note that this didn't mean that they might not object in the future.

Outside the court, RIAA tried to make the entire lawsuit appear frivolous. Executive Cary Sherman issued a public statement calling the decision to sue the organization "inexplicable." Both RIAA and SDMI had "unequivocally and repeatedly" declared, he said, that they had no intention to sue Felten or his colleagues, and he accused the EFF of continuing the legal battle "to keep their publicity machine running."[39]

The EFF responded with a statement of its own claiming that the suit was important because "RIAA threatened the scientific process" and that lawyers for Verance and RIAA had spent weeks threatening litigation and demanding that the researchers change the paper. For the EFF, allowing the paper to be published as presented was important. But the EFF's real goal was "a clear legal determination that no one needs the permission of the record companies before publishing and presenting scientific work."[40]

Sherman and RIAA's lawyers convinced the court that there was no conflict to resolve. Despite their previous actions, they claimed to have no objections to the presentation and publication of the paper, so there was no immediate cause to complain. But the decision provided no reassurance that they wouldn't try to block future research, by Felten's team or anyone else. This was a perfectly legitimate legal outcome that completely overlooked the root issue.

The content of the science was now inseparable from its legal history. Felten, USENIX, and the EFF webcast a presentation of one of the papers and followed it up with a panel on SDMI and the DMCA.[41] The case was triggering a wave of opposition to the DMCA among scientists, who saw their work threatened by the prospect of censorship masked as copyright protection.

Court documents and press releases argued that the future of the scientific method itself was at stake. Felten declared: "I understand that [RIAA and Verance] advocate an interpretation of the DMCA that would outlaw analysis of systems that might be used to control the use of copyrighted materials. However, I can say that such an interpretation would effectively prevent analysis of critical systems, and so would have a disastrous effect on education, research, and practice in computer security."[42]

In a court statement quoted by the Electronic Frontier Foundation, Felten explained that this would threaten not just computer science but all fields where the scientific method is important, since scientific inquiry requires "skeptical analysis of technical systems made by others, and the presentation of detailed evidence to support such an analysis." To outlaw this kind of analysis, Felten argued, "is to outlaw the scientific method itself."[43]

I often run into legal invocations of the scientific method. In his elegant book *The Scientific Method,* historian Henry Cowles showed how despite the popular belief that there is a single scientific method, there have been many variations of it since the nineteenth century, each built on a different way of imagining what it means to control nature.[44] But Felten's version of the scientific method was not about nature. It was about the Oracle. The scientific method, in his case, was a long process of trial and error to determine what it took to beat a black box of code.

Still, like so many other invocations of the scientific method in legal settings, this one was an effort to talk about systematic inquiry as a fundamental social need. Felten's statement advanced the myth of a unitary scientific method to warn that industrial hunger for copyright enforcement was threatening science itself. The stakes were not just new knowledge but the very act of doing science.

What Felten's statement left out, but was clear to programmers around the world, was that the issue was much bigger than that. The anticircumvention provisions of the DMCA said nothing about the scientific method or communities of scientists. But computer scientists and hackers were very specialized communities. Everyone was a potential target, and few in that community had Felten's resources to fight back.

This became evident in the summer of 2001, when the FBI arrested a Russian computer scientist for cracking a file security system. In a few weeks, scientists and hackers around the world would learn that the DMCA could lead to real criminal charges, millions of dollars in fines, and boycotts of the United States among international researchers. At stake were not just millions of dollars in the electronic publishing industry but researchers' rights to investigate proprietary technologies and share knowledge internationally.

FREE DMITRY!

In July 2001, thousands of hackers from around the world arrived in Las Vegas for the ninth annual DEF CON, one of the most popular hacking conferences in the world. DEF CON was known globally as a "computer underground party" where hackers could meet each other, learn about cutting-edge developments, or "just hack the network." Participants were instructed to bring everything they would need for a party in a desert city—from network gear and handheld transceivers to sunscreen and "what you would wear to a rave."[45] DEF CON was about computer science and about the fun of hacking. It featured many social events. There were coding competitions during coffee breaks and a race to find celebrity telephone numbers. The race's organizers, known as DrunkenWhores.com, promised to "try very hard to keep it legal, but still very fun."[46]

This is what brought Dmitry Sklyarov, a PhD student from Moscow, to the United States. He was scheduled to give a talk titled "eBooks Security—Theory and Practice," in which he would discuss the "security aspects of electronic books and documents" and demonstrate "how weak they are."[47] His talk would cover PDF encryption, several password recovery tools for digital documents, and password vulnerabilities in PDFs, specifically with Adobe's e-book reader.

This was not Sklyarov's first project on digital rights circumvention. He was building on previous presentations in which he had shown how "encryption in Microsoft Office documents is very weak and password protection may be removed without any problems in most cases."[48] At DEF CON, he was eager to present a program called Advanced eBook Processor, which would allow people to transform e-books in any format into PDFs.

Sklyarov had developed these programs while working for ElcomSoft, a Russian firm created in 1990 and a member of the Russian Cryptology Association. Its most popular product was Password Recovery Software, a tool the company advertised as essential for "businesses to continue to use their valuable data when documents' passwords are lost, accidentally or intentionally." It was designed for use with Microsoft Office, Lotus SmartSuite, and Adobe Acrobat's password encryption tools.

ElcomSoft also made other useful products, including Advanced Registry Tracer, which took snapshots of the Windows registry "to recover from systems crashes," and Advanced Disk Catalog, which quickly made a catalog of all media contained in a hard drive. All software was available for purchase online, and some of it had won awards from organizations such as ZDNet and TUCOW.[49]

ElcomSoft's "About" page described Advanced eBook Processor, or AEBPR, in very simple terms: "The only thing the program does is: converting documents from Acrobat eBook format (compiled for Adobe Acrobat eBook Reader) to the plain Acrobat format (PDF). Again, that's all: from one Adobe format to another." The website also had a legal caveat: "This program works only with eBooks you legally own, i.e. purchased from . . . online stores like Amazon or Barnes & Noble. So we were absolutely sure that the owner of eBook has all rights to read the book he purchased where he wants and how he wants."[50]

I often saw similar notes in the late 1990s and early 2000s when I downloaded illegal digital versions of video games. I had a collection of special programs called emulators that allowed me to play any console's games on my PC. Emulators' pages sometimes indicated that US law allows users to make one copy of a game for archival purposes and that the emulator and digital files were meant to serve only that purpose. No one I knew used them for that.

Back at DEF CON, Sklyarov was about to enter a legal war zone. A few days after ElcomSoft released AEBPR, a man named Daryl Spano, a technical investigator at Adobe's Anti-Piracy Unit, found out about it in an online forum. The program was what is called "shareware," meaning that users could download it for free and use it with limited capabilities. If they wanted the full program, they could purchase a digital key to unlock it.

The free version of AEBPR let Spano see about 10 percent of an e-book and instructed him to buy a license for $99 if he wanted to unlock the full program. The program's opening screen also named Dmitry Sklyarov as the copyright holder. Purchasing the license was an international endeavor. A US company called Register Now! would collect the payment and forward the information and money to ElcomSoft in Russia. Then the company's managing director, Vladimir Katalov, would email the unlocking key to the user.[51]

On its website, ElcomSoft alleged that Adobe's strategy was to target both its ability to keep the program available for download and the international systems that allowed it to receive payments. Spano's team emailed ElcomSoft to say that AEBPR was illegal and that it must be immediately removed from the company's website. Adobe also contacted ElcomSoft's Internet service provider, a company called Verio, to request that the company's website be shut down because it offered downloads of Adobe's copyrighted software. I was unable to verify this based on court records, but if, in fact, ElcomSoft was not hosting Adobe software, then Adobe's alleged request to Verio would have been grounded on a lie.[52]

Verio nonetheless blocked all access to ElcomSoft's website from its own servers. ElcomSoft set up mirror websites on various servers to ensure users' continued access. Its program was still available, so Adobe targeted ElcomSoft's finances by complaining to Register Now!, which responded by cutting ties with ElcomSoft.[53] If ElcomSoft had then arranged with a new company to process payments, Adobe would have probably complained to them, too.

Adobe's actions caused ElcomSoft to go on the offensive. In early July, it published a seething critique of Adobe's security protocols: "Now it's time for the brutal truth on Adobe eBook protection. We claim that ANY eBook protection, based on Acrobat PDF format (as Adobe eBook Reader is), is ABSOLUTELY insecure due to the nature of this format and encryption system developed by Adobe."[54]

ElcomSoft's post explained that the "general rule" to bypass e-book protection was to convert the e-book file to a plain PDF, adding that not a lot of experience was needed to do so. Adobe's security plug-ins could involve hardware verifications, connections to Adobe's websites, and several encryption methods, but all these protections were simply designed to send a decryption key to Acrobat readers. Users who had one of these keys—for example, by having a copy of AEBPR—could force the files open. According to ElcomSoft, Adobe had violated authors' rights by "aggressively pushing" such simple encryption methods.[55]

ADOBE AND THE FBI

Adobe contacted the FBI to report that ElcomSoft was violating the law.[56] Daniel J. O'Connell, a veteran FBI agent assigned to the High-Tech Squad in San Jose, California, met with two Adobe employees: Kevin Nathanson (group products manager for e-books), and Daryl Spano. They showed O'Connell assorted literature about AEBPR that described how the program creates a "naked file," allowing people to read an e-book on any computer without having to pay a fee to the bookseller. They also told the agent that Adobe was being "victimized" by ElcomSoft.[57]

O'Connell started his own investigation in early July. To verify Adobe's reports, he browsed ElcomSoft's website and Register Now!'s listing. He also noticed that Sklyarov was scheduled to speak at DEF CON. Sklyarov claimed copyrights over the program and was bringing it with him to the United States. He was going to explain to others how it worked, and he might even be distributing copies of it.

Before doing anything, O'Connell needed to check a few technical facts. He called Tom Diaz, a senior engineering manager for e-books at Adobe. In Diaz's view, the program did this "by avoiding, bypassing, removing, deactivating, or otherwise impairing the technological measure." This was exactly what O'Connell needed to hear. He concluded that Sklyarov had "willfully and for financial gain imported, offered to the public, provided, and otherwise trafficked in a technology, product, service, and device" design to circumvent a technology protected by the Digital Millennium Copyright Act.[58]

On July 16, the FBI arrested Sklyarov, and all signs suggested that Adobe would make an example of him. He was charged with trafficking and distributing a program that could circumvent technological copyright protections and with aiding and abetting ElcomSoft. He remained in jail for about three weeks, when he was released on a $50,000 bail and ordered to stay in Northern California.[59]

The Department of Justice issued a press release calling this "one of the first prosecutions" under the DMCA.[60] Citing DEF CON's website, the release explained that Sklyarov had come to the United States to participate in a "computer underground party for hackers" where he would discuss "security aspects of electronic books and documents." The maximum penalty for the charges against him was five years in prison and a $500,000 fine. Sklyarov was detained without bail and relocated to the Northern District of California, where Adobe had filed its complaint. The case would be prosecuted by a top Justice Department team: attorneys Scott Frewing and Joseph Sullivan would prosecute the case, assisted by Lauri Gomez.[61]

The EFF, its memories of Felten's struggles still fresh, took an immediate interest in the case. Shari Steele, its executive director, charged that "Adobe, seeking to protect electronic property rights at any cost, has apparently pushed the U.S. Department of Justice into an ill-advised arrest."[62]

What bothered Steele was that the case was neither about copyright infringement by Sklyarov nor about his own use of the program. It was about the act of *making* the program. That is what the DMCA had made illegal.[63]

AGAINST ADOBE

Digital rights activists quickly took matters into their own hands. Robin D. Gross at the EFF stated that the foundation would join "the international community in expressing outrage." Allies protested the "disgraceful arrest" and "grave miscarriage of justice" against Sklyarov. Rallied by the EFF, these activists would take the fight to the streets and called for a boycott of Adobe's products.[64] They had an interesting logo that I cannot show here for copyright reasons but which you can see for yourself at the link in the notes.[65]

On July 23, small crowds gathered in Moscow and twenty-one US cities to protest Adobe and Sklyarov's imprisonment. Protesters in San Jose, California—Adobe's backyard—marched for two hours holding up Russian flags and homemade signs. Because the EFF had suggested that signs have four words or fewer "so that they are easy to read for people passing by," they carried slogans such as "LEGALIZE PROGRAMMING," "PROMISE SECURITY, DELIVER LIES," "CODING IS NOT A CRIME," "FREE SPEECH FREE DMITRY," and "ROT13 IS NOT A CRIME" (a nod to ROT-13, a simple letter substitution code).[66] You can see photographs of these protests at the link in the notes, since I was unable to obtain permission for their publication in a book.[67]

Overseen by mounted Bay Area police, men and women marched while carrying pictures of Sklyarov and his young family.[68] In Boston, groups of forty to a hundred protesters carrying signs such as "GEEKS 4 FREE SPEECH" organized at local restaurants and across the Charles River at Harvard and MIT.[69] They presented petitions to Senator Ted Kennedy and marched to the Boston Public Library, where they connected with activists from the Coalition for Free Speech.[70] Protesters handed out flyers, spoke with major outlets such as the BBC, and reworked popular songs into protest chants.[71] One group, inspired by the Beastie Boys, chanted, "You gotta fight, for your right, to reeeeaaad BOOKS."[72]

In Moscow, about ten people gathered at the American embassy, handing out flyers to people standing in line at the embassy or walking by. You can see photos from the event at the link in the notes.[73] An organizer named Ilya Vasilyev, from the Civil Hackers' School, reported: "Flyers were specially designed to tell those people about one Russian, that already visited that Embassy, wishing to fly to America, but don't even realized, what it will cost to him. Flyers have two pictures of Dmitry—at 15 of July, 2001 he is respectable inter-tourist in U.S., scientist. At 16 of July, 2001 FBI dressed him, like they used to dress a criminal types."[74]

According to Vasilyev, handing out flyers was his group's way of informing a thousand Russians who wanted to visit the United States "that they can be put there in prison by FBI for committing no crime" and that leading scientists were withdrawing from US conferences to protest Dmitry's treatment. Vasilyev reported: "We hold two flags—U.S. and Russian, different signs and 'FREE SKLYAROV' message on two languages. My beloved sign 'SVOBODA MYSLI—SVOBODA CODA' ('FREE THOUGHT—FREE CODE') was taken by a girl Ivy, my student, in order to nail to the wall of her own room."[75]

The Moscow protesters were a small but surprisingly active crowd. They organized an e-protest of Adobe and the FBI, where they noticed that a few hackers "really wanted their blood." They insisted that their digital activism "was completely legal and fine."[76] A few protesters wore black "for mourning of human rights in U.S. and Internet."[77]

Sklyarov and ElcomSoft were indicted on August 28, 2001, on four counts of circumvention and one count of conspiracy to traffic a program capable of circumventing copyright protections. Sklyarov faced up to twenty-five years in prison and more than $2 million in fines; ElcomSoft faced about $2.5 million in fines. Sklyarov was released from US custody and allowed to go back to Russia that December after he agreed to testify against ElcomSoft.[78]

The next year, to most observers' great surprise, a federal jury found Sklyarov not guilty on all counts. We will probably never know what happened inside the jury room, but what may have tipped the scale, according to media reports, was that the judge had instructed the jury to issue a guilty verdict if they found that ElcomSoft and Sklyarov had the intention of violating the DMCA. This jury, at least, was not convinced.[79]

The point here is not whether Sklyarov was found guilty. Simply charging him was enough to derail his life for more than a year. He was kept in US custody away from his family and underwent a very stressful ordeal simply because he had developed a computer program. Like Princeton's Felten, he was experiencing a new way in which copyright could become a censorship tool. Both men got in trouble for investigating a problem, creating a method to solve it, and then talking about it with a broader audience. Under the DMCA, copyright was becoming a tool of censorship even without direct infringement of someone else's rights to distribute or duplicate a creative work.

©©©

Despite the DMCA's anticircumvention provisions, it is very easy to find "cracked" software if you know where to look. I'll tell you about one website where this is easy to do. However, to protect myself from libel allegations, I will use a made-up impossible URL (C*N.com) and will not quote the website verbatim or mention exact prices. If there are any serious gamers in your life, I'll bet they can help you find many sites like this. The website works and it is very popular—my students and several people I know use it to get video games for their PCs without having to pay full price.

C*N.com looks like any other online video game store. When I last visited it, on May 24, 2024, its main page advertised a Memorial Day Sale and had several high-quality images. The site feels like a marketplace for used books in that it does not appear to sell anything directly. Instead, users make their own online storefronts. Each item's listing looks like a video game, but what sellers are offering are "keys": the codes you enter into a game or some software after installing it so that it can run on your computer. If you make a purchase (using any major credit card or PayPal), the seller will send you the key for your game so that you can download and play it. The crucial step here is that sellers offer keys for a legitimate gaming platform called Steam, where many popular PC games can be found.

I'm sure someone has argued that C*N.com is designed for a perfectly legitimate purpose: if you buy the key to unlock a video game but then change your mind, you could sell it at C*N.com. But that is not what happens. Instead, sellers can generate keys using special hacking software, and their clients use special game unlocking software to crack the game. For example, I could buy a key for *Red Dead Redemption 2* for less than $25, instead of almost $70 if I buy a copy at Target. Many sellers offer this key at C*N.com, but I would probably go with one that has made more than seventeen thousand sales and has 98 percent positive feedback.

For an American game maker, going after C*N.com and its sellers would be nearly impossible. The URL is registered in the United States, but the C*N.com's work is spread all around the world. It has a US office, but that appears to be more about marketing than anything else. Instead, judging from its extensive list of job openings, all technical work takes place in East Asia and at two European companies. And even if US companies could get a judge to shut down C*N.com, the sellers are scattered throughout the world, and it would only be a matter of time before another third-party website brings them all together again.

CHAPTER FIVE

Broken Voting Machines

IN THE SPRING OF 2023, news broke with allegations that fake DMCA takedown notices (and their European counterparts) were part of a major global disinformation campaign led by governments, drug cartels, and other powerful organizations. The accused companies were dedicated to erasing content from the Internet and listed fake DMCA notices as part of their portfolio of services. We learned these things thanks to a French nonprofit agency called Forbidden Stories, which continues work started by journalists who would be endangered if they pursued their stories any further.

One of these firms was Eliminalia, a Spanish company founded in 2013 by a man named Diego "Didac" Sánchez. Forbidden Stories obtained leaked files suggesting that Eliminalia had worked with fifteen hundred individuals and corporations across fifty countries. The firm's clients allegedly included a British slumlord, a Swiss bank accused of money laundering, a Turkish millionaire accused of murder, and a Mexican drug cartel. Eliminalia reportedly offered clients the ability to control the spread of news stories and information about them and helped them clean up their online personas. Allegedly, if a client wanted a news article deleted, Eliminalia would find a way of getting it done, making it, in essence, censorship on demand.[1]

According to Forbidden Stories, Eliminalia employed a varied portfolio of tactics, including use of the European "right to be forgotten" laws.

Grounded in the European Union's General Data Protection Regulation, these laws grant people the right to demand that an organization delete its personal data about them. For example, the European Court of Justice sided with a man who demanded that his debts be removed from Google Spain. (The legal scholar Meg Leta Jones has written a fantastic book on these laws titled *Ctrl+Z*.)[2]

Forbidden Stories also accused Eliminalia of using copyright law to remove content from the web. It allegedly employed the back-dating technique I mentioned earlier—creating a clone of a page, setting a publication date prior to the original, and demanding a DMCA takedown. Forbidden Stories alleged that the company marked a targeted site as "eliminated" after it filed a DMCA complaint. Automattic, the parent company of WordPress, the popular webpage design and hosting service, acknowledged to the *Washington Post* that there is "a lot of potential for abuse as a way to censor speech and legitimate criticism."[3]

Unfortunately, we may never know more about Eliminalia than what Forgotten Stories found from the leaked documents. Eliminalia's contracts allegedly require both employees and clients to operate under strict confidentiality. In 2023, the company was renamed Idata Protection. We may not be able to zoom into Eliminalia, but we can zoom out and think about how other corporations might deploy copyrights as a form of censorship.

I became obsessed with the problem of corporations deploying copyrights as a form of censorship back in 2020, when I was a national fellow at the Hoover Institution at Stanford. While investigating the relationships among commerce, democracy, and copyright law, I ran into the story of a manufacturer of voting machines called Diebold. During the US presidential primaries in 2004, this company had inflamed the growing mistrust in our national electoral infrastructure by manufacturing flawed machines—and then appearing to censor folks who talked about them.

What I found fascinating about Diebold's story is that the firm actually owned the copyrights over the materials it was trying to delete. An enormous collection of company files had been leaked to the public and was spreading like wildfire on the eve of a contentious presidential primary. Faced with few ways to contain the leak, Diebold turned to copyright law in an effort to challenge a popular adage of the time: that information *wants* to be free.

THE DIEBOLD DEBACLE

During their first year at Swarthmore College, in the spring of 2003, undergraduates Luke Smith and Nelson Pavlosky craved a community where they could think about alternatives for traditional intellectual property protections. In their words: "We lamented the lack of an organization to spread the ideals of the open source / Free Software movement to other kinds of so-called 'intellectual property.' "[4]

The notion that human creations could be "free" was gaining considerable traction among legal scholars and some computing professionals. For example, the free software movement, founded in the mid-1980s by Richard Stallman, proposed that users should be free to use, share, study, and modify software however they wish.[5] The word "free" in free software does not mean that users don't pay for the software. It's an ideal. To quote a popular biography of Stallman, the movement revolves around the notion of "free as in freedom."[6] Christopher M. Kelty's excellent book *Two Bits* shows how the philosophy of free software extended beyond software and into other realms of media and knowledge.[7]

By the time Smith and Pavlosky got to Swarthmore, the free software movement was part of a broader range of efforts to delineate alternatives to traditional copyright protection that favored the free circulation of creative works and ideas. Chief among them was Creative Commons, an organization created in 2001 and led by a stellar group of legal scholars, artists, and entrepreneurs, including Lawrence Lessig. Their goal was to create a range of licenses that allowed creators to set the conditions under which others were allowed to use their work (figure 11). According to the earliest version of the Creative Commons website available for research (2003): "We use private rights to create public goods: creative works set free for certain uses. Like the free software and open-source movements, our ends are cooperative and community-minded, but our means are voluntary and libertarian. We work to offer creators a best-of-both-worlds way to protect their works while encouraging certain uses of them—to declare 'some rights reserved.' "[8]

Smith and Pavlosky knew that Swarthmore's student culture was fertile ground for the ideology of "freeness" to take hold. The campus's politics have historically leaned liberal. The college also has a very active student-led information technology group called the Student College Computer Society. Founded in 1992, the SCCS was in charge of providing email, web space, and UNIX accounts to the college's faculty, students, and staff. On the tenth

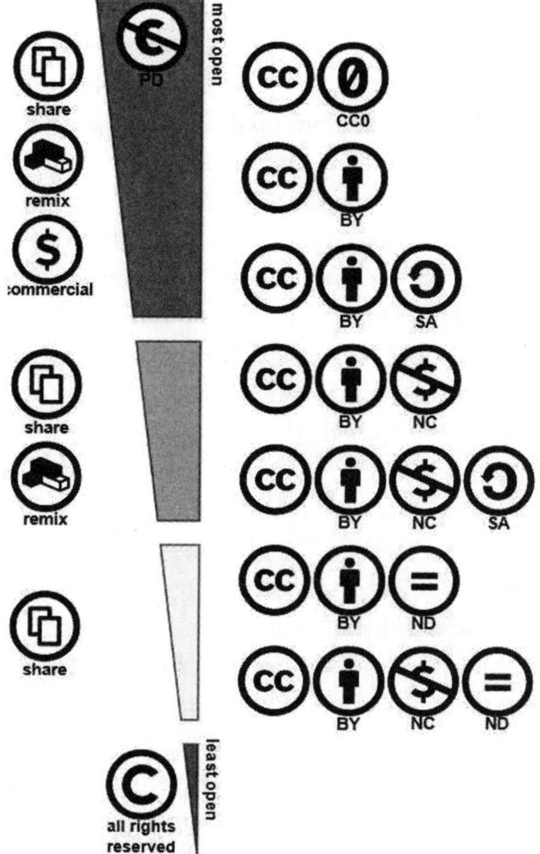

FIGURE 11. You've been seeing the influence of Creative Commons (CC) in the image captions throughout this book. The right-hand side of this chart arranges the different types of CC licenses from most restrictive (at the bottom) to least restrictive (at the top). Each line is a different kind of license, and each circular logo corresponds to a different restriction. For instance, the circle with a person means "Attribution" and requires the user to credit the author. This image was uploaded to Wikimedia Commons by user Sintegrity, and I modified it only to increase its resolution. It is licensed under the Creative Commons Attribution-Share Alike 4.0 International License (https://creativecommons.org/licenses/by-sa/4.0/deed.en).

anniversary of its founding, it opened a SCCS media lounge to the entire Swarthmore community, providing both access to computing technologies and a space for folks to meet.[9]

Swarthmore had no student group dedicated to free software, so in a flash of what they called "freshman year idealism," Smith and Pavlosky decided to create one and called it the Swarthmore Coalition for the Digital Commons (SCDC).[10] It was as informal as a student club could be, and it soon fizzled out. As they would recall later, "It was not easy going, that year. It was already the middle of the spring semester, and few people came to our largely incoherent meetings. Worse, they weren't the right people—they had no commitment to the cause; they were more interested than involved. The school year ended, and that was that."[11]

The two students decided to try again as sophomores. They lured first-year students into their meetings with pizza they paid for themselves and gave presentations based on Lawrence Lessig's book *Free Culture.*[12] Lessig's ideas allowed Smith and Pavlosky to make the same kind of leap that Kelty described in his book: they took the broader philosophy of what it means to be "free" and applied it to realms other than software. Smith and Pavlosky hoped to use the SCDC to host all kinds of events, from a LAN party to meetings with university administrators. But they found their organization's first mission thanks to an unexpected announcement in July 2003: a publicist and writer named Bev Harris had obtained forty thousand files belonging to a maker of vote-tallying machines, Diebold Election Systems.[13]

Harris published her story on a New Zealand–based server, perhaps to keep it out of US courts' reach. Her most troubling revelation was that it was very easy to change the final vote count in a Diebold machine. A person could easily add or subtract votes manually without the system raising any red flags. Even audit trails could be altered without generating any system warnings, and passwords protecting administrative access could be bypassed. Yet Harris didn't just *say* she had done these things. She posted step-by-step guides with system screenshots to show users how to do it, too. But there was much more to the files than that.[14]

THE DIEBOLD FILES

Harris found these files during a very low point in the US public's relationship with voting technology. George W. Bush's historically narrow victory over Al Gore in the presidential election just two years earlier had been contentious because tabulating machines had disqualified millions of ballots. Machines in Florida had missed counting sixty-one thousand votes. The state's supreme court allowed a recount, but the US Supreme Court stepped in with a 5–4 opinion prohibiting the tabulation. This judicially validated miscount gave Bush the victory even though Gore had won the popular vote.[15]

Since then, President Bush had signed a new voting technology law, the Help America Vote Act of 2002.[16] This law replaced older voting machines and created new agencies and standards to make the election process smoother. Voting machine companies started to compete for contracts at the local, state, and federal levels, and one of the winners in these races was Diebold, whose systems had already been used in several elections. There was potential for enormous profit, but only if companies could reach voting districts quickly, convince local leaders that their machines were the best in the business, and cut costs wherever possible.

The Diebold files cast further doubt on Bush's victory and suggested that things had not improved much ever since. How they got online is still a mystery, though I suspect that a whistleblower inside the company took the initiative to compile and release them. Harris reported that she found them on a public website known to industry insiders but not to public interest groups. Some of the files must have been trade secrets. They included diagrams of communications setups, source code, manuals, passwords, encryption keys, simulators, and even "files loaded with votes and voting machine software." Taken together, Harris wrote, the files were "a virtual handbook for vote-tampering."[17]

Some messages suggested that company employees were willing to pretend that the machines met local requirements. One message sent in July 1999 by a sender I'll call John Dee read: "We, the manufacturer, are supposed to set the procedures to follow for this equipment since we build it. I hate more than anyone else in the company to bring up a certification issue with this, but a number of jurisdictions require a 'system test' before every election. . . . That is why the AccuVote displays the silly ***System Test Passed*** message on boot up instead of 'memory test passed,' which is all it actually tests. No argument from me that this is pointless. You could probably get away with a batch file that prints 'system test passed' for all I know. We will do something along those lines with the new unit after a memory test or whatever."[18] (It is, of course, impossible to tell whether Dee was the one who sent this message.)

Emails also indicated that the machines had produced such outrageous election results in 2000 that local leaders were scrambling to face auditors. A message sent in January 2001 from a client in Florida read: "I need some answers! Our department is being audited by the County. I have been waiting for someone to give me an explanation as to why Precinct 216 gave Al Gore a minus 16022 when it was uploaded. Will someone please explain this so that I have the information to give the auditor instead of standing here 'looking dumb.' "[19] This sparked an email exchange among employees trying to figure out how this could happen. At one point one of them asked colleagues "to shed some light here, keeping in mind that the boogie man may [be] reading our mail."[20]

With a new election on the horizon, the files suggested that some employees were interpreting districts' dissatisfaction with voting machines as opportunities for profit. For example, in January 2003, likely after public statements documenting dissatisfaction with the machines, one email sent on Dee's behalf read: "There is an important point that seems to be missed by all these articles: they already bought the system. At this point they are just closing the barn door. Let's just hope that as a company we are smart enough to charge out the yin if they try to change the rules now and legislate voter receipts."[21] A follow-up message sent from Dee's account clarified what "yin" meant: "Short for 'out the yin-yang.' Perhaps a little too colloquial; apologies for that. In my defense, google turns up 694 references to the phrase. Any after-sale changes should be prohibitively expensive. Much more expensive than, for example, a university research grant."[22]

FREEING THE FILES

Just days after Harris published her blog post, news outlets and media commentators were discussing the files and activists around the country were reposting them on their servers.[23] One of these groups was called Why War?, an antiwar nonprofit that Swarthmore students had created in the aftermath of the September 11 attacks. The group posted the emails on an off-campus server, created a list of links to other sites that hosted the emails, and asked readers to post them in their own newsgroups. Like all other groups hosting these files, Why War? came under attack. Diebold sent out a flurry of complaints to online server providers, demanding that the files be taken down and threatening legal action.

A month or so after the story broke, Smith and Pavlosky concluded that this was the kind of issue the Swarthmore Coalition for the Digital Commons needed to care about, and they decided to place the email archives on the SCDC website, which was hosted on Swarthmore College's servers.[24] In doing this they were simply following the ideology, popular among hackers, that "information wants to be free." Historian Adrian John's book *Piracy* shows how this sentiment has taken on many meanings over the years, from critiques of corporate and legal control over information technology to speculations about the economics of information storage and flow.[25] Smith and Pavlosky were following Lessig's *Free Culture* approach, in which Internet technologies are mobilized as a democratizing force that grants access to information.

Robert Gross, the dean of the college at Swarthmore, called the two students into his office the day after they posted the emails. He explained that Judy Downing, the university's director of information technology services, had received a letter from a Diebold intellectual property attorney named Ralph Jocke. He threatened a lawsuit over copyright infringement and demanded that Swarthmore "remove and destroy the Diebold Property" now on the college's servers.[26]

Gross was doing what he could to avoid a legal conflict. The letter was a few weeks old—an individual student had posted the email archive to a Swarthmore dormitory server right after Harris's story came out—and Downing's office had identified many other instances as well. Gross called a public meeting to explain that Swarthmore "can't get out in front in this fight against Diebold."[27] More than fifty people attended, and a few hundred more emailed him to express their views. The college was also moving quickly to comply with Diebold's request: it disabled access to the emails hosted in the SCDC website and even prohibited Smith and Pavlosky from including a link to email archives in servers outside Swarthmore.[28]

Why War? and the SCDC were deeply disappointed. They saw the college's response as an affront to both their politics of information and Swarthmore's values as a liberal arts school. "We had hoped," wrote a Why War? member, "that an institution once praised for allegiance to the pursuit of truth would have taken a more forceful stance in defense of information."[29]

The students decided that it was time for "electronic civil disobedience."[30] The plan was for Why War? and the SCDC to work in parallel. Why

War? would focus on ensuring continuous access to the archives by publishing mirror websites in different servers (maybe some abroad) and linking to all of them in its main page. The group could no longer host the files on its own site, as the website's host had received a similar letter from Diebold and decided to take the content down.[31] The SCDC, meanwhile, would create a website outside the Swarthmore network and fight as hard as possible to keep the archive available there.

The Swarthmore administration doubled down on its enforcement of Diebold's demands. Students were warned that placing any link to why-war.com from a university-hosted website would be a punishable offense. At Why War?, where years of activism against the Patriot Act of 2001 had made free speech into a fundamental organizational value, this was egregious. Why should students be punished for merely saying where information was available? Was it now a punishable offense at Swarthmore to spread knowledge? The students saw creating these hyperlinks as an act of civil disobedience in the service of free speech.

Swarthmore's ban on linking was probably beyond its legal duty, but online records and court documents suggest that the university was making a good faith effort to comply with DMCA's requirements while still listening to the students' concerns. Soon after enforcing their stringent rules, university officials told the SCDC that the DMCA allowed for a counter-notification process. This is a formal legal response that students could file in response to Diebold's takedown request, which would require the university to reenable access to the archives at least until Diebold contested the notice.[32] Downing agreed to receive such a notice, and Smith and Pavlosky wasted no time in filing it. They declared, under penalty of perjury, that they had a "good faith belief that the Diebold email archive was disabled in error as a result of mistake or misidentification of the material, including that our posting of this material constituted fair use under copyright law."[33]

DIEBOLD VERSUS THE INTERNET

Diebold's lawyers soon realized that it is almost impossible to make something disappear from the Internet. Media outlets such as *Philadelphia Daily News*, *Wired*, and the *Inquirer* were reporting on the company's takedown notices. Several reports portrayed the SCDC and its allies in digital activism as a surprisingly coordinated coalition of Davids fighting a corporate Goliath.[34] Students elsewhere started posting the materials on their own universities' servers, including those at MIT, Purdue, the University of Texas at Austin, and the University of Southern California.

But the Help America Vote Act had made it an unusually profitable time to be a voting machine manufacturer, and Diebold was not slowing down. Ralph Jocke's office was very busy sending out takedown notices on Diebold's behalf, while company spokespeople issued statements decrying the act of posting the materials online to be copyright infringement. An article in the *Chronicle of Higher Education*, by a geologist named Andrea Foster, reported that one spokesperson emphasized that the company "would continue to send copyright infringement notices to Internet service providers that host the company documents."[35]

As Diebold followed up on this promise, website after website took down the email archives. Even Bev Harris had her site taken down after her website's host received the company's letter.[36] A handful of groups resisted the takedown notices, including the SCDC and Indybay, the San Francisco branch of an independent media collective called Indymedia that was dedicated to promoting online free speech and Internet access for underserved populations.

The email archive's continued accessibility through Indibay's links showed just how hard it would be for Diebold to make the content disappear. Users could post comments on the website's news items, including the page about the archive. Every time Diebold shut down the servers listed in the original post and the comments, a user would post a new comment with a link to a new server that hosted the full archive. Users shared text-searchable databases and the full texts of some of the emails, and they discussed creating a newsgroup where everything could be shared directly into people's inboxes.[37]

Students and activists were determined to ensure that the emails would survive Diebold's attacks. Smith told *Wired* that the plan was to bypass the company's threats by creating a chain of hosting duties. Whenever Diebold shut down a machine or server, the SCDC would arrange for a different machine to start hosting the archive. "They're using copyright law as a means of suppressing information that needs to be public," Smith told *Wired*. "It's a great example of how copyright law can be against the public good rather than for it, as it was originally intended." No one was looking at the emails to "steal Diebold's election system." The company was attempting "to conceal flaws that directly affect the validity of election results."[38]

By early November 2003, with the presidential primaries just a few months away, Diebold's struggles to remove its internal emails from the Internet had become national news. Speaking to the *New York Times*, Rebecca Mercuri, a computer scientist at Bryn Mawr best known in legal circles as for her expert testimony in *Bush v. Gore,* asked, "Are these companies staffed by folks completely ignorant of computer security . . . or are they just blatantly flaunting that they can breach every possible rule of protocol and still sell voting machines everywhere with impunity?"[39]

At Diebold, academic criticism was far less important than the damage to the company's reputation and profits. In California, for example, the assistant secretary of state for policy and planning announced that the state would halt certification of Diebold's machines because of "disconcerting information."[40] One reporter wrote that Diebold representatives sitting in the audience "seemed surprised by the announcement" and "expressed displeasure" that the decision had been announced so publicly.[41]

Faced with this multifaceted crisis, Diebold doubled down on its efforts to make the emails disappear while holding tightly to the assertion that it was just trying to protect its intellectual property. If it could get a court to rule that the presence of the emails in an Internet service constituted copyright infringement, then the company would be able to cite that opinion as precedent in its cease-and-desist letters. More important, such a precedent could deter other website hosts from allowing users to upload the emails, and it would provide a powerful weapon with which to counter groups such as the SCDC, Why War?, and Indybay, where hosting the digital archives amounted to free speech activism.

COPYRIGHTS AS CENSORSHIP

The activists' resistance gained legal momentum when the EFF and the Center for Internet and Society at Stanford Law School took an interest in the case. Diebold had sent a takedown notice to the Online Policy Group (OPG), which hosted Indybay's website and provided the organization with Internet access. At the EFF, Wendy Seltzer was ready to lead a defense of OPG in the hope of creating precedent that would prevent the weaponization of copyright law as a censorship tool. Seltzer is one of US copyright's superstar lawyers, having risen to national prominence through her creation of the Chilling Effects Clearinghouse, a joint project involving the EFF and law clinics across the country.[42] She was determined to fight back against the idea that simply linking to content could be copyright infringement. In October 2003, when the email controversy was starting to enter national media cycles, Seltzer answered Jocke's takedown notice with the explanation that OPG "declines to remove the IndyMedia pages" with the links to the material because a person cannot infringe on a copyright merely by linking to content. Linking, she wrote, "is not among the exclusive rights granted by the Copyright Act."[43]

Seltzer also pushed fair use and free speech defenses. She insisted that "the postings themselves are plainly fair use, not infringement." Indymedia "is a new organization whose use of these links gives background to its discussion of the controversy surrounding e-voting." She explained that the First Amendment "plainly protects" OPG's pages, but her argument was mostly a copyright defense: "Even if Diebold has an enforceable copyright in the documents, their reposting by others serves the public interest and would be deemed fair and non-infringing."[44]

By October 2003, Diebold had not yet filed any lawsuits, but it was continuing to send cease-and-desist letters across the country, so the EFF coordinated a bolder effort: it requested a temporary restraining order that would bar Diebold from sending any more cease-and-desist letters or take-down notices. It was unclear how far the company was willing to go. It was unlikely to succeed in removing the content from the web, but there was some danger that it would get a court to agree that linking to the emails was copyright infringement.

To get the restraining order, Seltzer joined forces with two more attorneys. The first was her superior at the EFF, Cindy Cohn, the legal director. Cohn had handled an extraordinary range of issues at the EFF and had been the lead attorney in *Bernstein v. US Department of Justice,* which established that a government restriction of computer code requires analysis under the First Amendment.[45]

The second attorney was Jennifer Granick, the clinical director at the Stanford Center for Internet and Society, which at the time was headed by Lawrence Lessig. A recent graduate of Stanford Law, Granick had become a prolific scholar of privacy, trademark, and security. She had practiced criminal defense in cases related to trade secret theft, email interception, and unauthorized access. She still handled cases as clinical director and would represent Smith and Pavlosky, who agreed to join the suit as plaintiffs.[46]

The three lawyers' filing for a temporary restraining order expanded on Seltzer's reasoning that Diebold was misusing its copyrights and making false claims under the DMCA. Specifically, the company had misrepresented "that fair use reproduction of the e-mail archive was infringing" so that it could "induce the takedown" of the archive and "take advantage of the DMCA's expeditious removal procedure." This was an especially grave charge, because the DMCA provides that a person who misrepresents that an activity or material is infringing is liable for damages, including attorney's costs to copyright owners and service providers.[47]

The filing, in the District Court for the Northern District of California, did not slow Diebold down. While the EFF lawyers were presenting their case to Judge Jeremy Fogel, Diebold's attorneys were sending their second letter to Hurricane Electric, an ISP that offered hosting and dedicated servers to OPG and other local access providers. Diebold's lawyers now argued that even hosting excerpts from the email archives in news items discussing the controversy was a copyright violation.[48] The letter, addressed to Benny Ng, Hurricane's marketing manager, implied very strongly that Diebold would not hesitate to sue Hurricane for all costs and damages associated with the files' online circulation. This was especially threatening given that entire states were reconsidering their relationships with Diebold on the eve of the 2004 election.[49]

In fighting the restraining order, Diebold did not budge from its position that this was entirely a matter of copyright infringement. Its lawyers argued that the plaintiffs sought to bar the company from exercising both its constitutional right to file a copyright infringement suit and its rights under the DMCA, which allowed it to "serve notifications of copyright infringement on online service providers." The lawyers, Robert Mittelstaedt and Adam Sand from the law firm Jones Day, were betting that the court would weigh copyrights more heavily than First Amendment rights.

Their primary legal weapon was the verdict in *Harper & Row Publishers v. Nation* (1985), where the Supreme Court had ruled against the *Nation* magazine for having published hundreds of words of excerpts from Gerald Ford's memoir before the book was released.[50] The Court found that fair use is not a defense against the prepublication of a work, even if its content is of substantial public interest. According to Mittelstaedt and Sand, this meant that First Amendment rights did not supersede copyrights and that public interest in the emails did not trump Diebold's rights to control their circulation.[51]

The most troublesome evidence about Diebold's systems was a technical report published in July 2003 by the Johns Hopkins University Information Security Institute. Researchers at UC San Diego, Rice, and Johns Hopkins had uncovered serious flaws in Diebold machines. Based on a security analysis of the system's source code, the researchers had shown that voters could cast unlimited votes without having any system privileges, that poll workers could modify votes, and that it was possible to "match votes with the voters who cast them."[52]

If law and research alone could not stop Diebold's lawyers, the court of public opinion could. The terrible publicity Diebold was generating would only worsen if the company started suing everyone, and the files were spreading like a digital wildfire as media outlets and prominent politicians continued to publicize them. Even a US representative, Dennis Kucinich of Ohio, had posted portions of Diebold emails in his website and was calling for Congress to investigate the company's abuses of the DMCA.[53]

On December 1, 2003, with the issue of the injunction still undecided, Diebold announced that it would not pursue legal action against anyone who had published its email archive and that it would send retractions of its legal threats to ISPs.[54] The company's president, Robert J. Urosevich, proclaimed that the withdrawal showed the company's "commitment not only to provide the best voting systems in America, but to contribute to a robust public debate on how to record and tally the vote most accurately and efficiently."[55]

Many activists saw this as a resounding victory for the creation of a fairer electoral system. On the SCDC's website, a celebratory website post linked to the entire archive and explained, "Diebold has withdrawn all of its DMCA notices, and has promised not to send us any more. They're very sorry, and they won't let it happen again. This means that you can mirror the memos freely! Nobody will do anything to you! We can now use these vital documents in public discussion without fear."[56]

Many organizations similarly focused on the election controversy surrounding the emails, not the copyright implications of Diebold's actions. At Why War?, the last update in a page dedicated to the controversy opened with "We won" and celebrated the public's continued ability to discuss election technology freely.[57] At the OPG, Director David Weekly told the EFF, "As an ISP committed to free speech, we are affirming our users' right to link information that's critical to the debate on the reliability of electronic voting machines."[58]

But the EFF was not content with Diebold's promise not to sue people, because that didn't resolve the bigger problems: how to ensure that copyright law was not used as a tool for online censorship and how to prevent future abuses of the DMCA's takedown notifications. "We're pleased that Diebold has retreated and the public is now free to continue its interrupted conversation over electronic voting machines," Seltzer wrote in an EFF media release. "We continue to seek a court order to protect posters, linkers, and ISPs who host them."[59]

To this end, the EFF lawyers joined forces with Alan Korn in their effort to obtain a restraining order. An art and entertainment lawyer specializing in free speech and intellectual property who had once been a radio disc jockey and recording artist, he had represented dozens of FM radio stations since graduating from law school in 1993.[60] He and EFF legal director Cindy Cohn would represent the Online Policy Group, and Granick would continue to represent the students.[61] The goal was to have a court rule that Diebold had abused the DMCA.

On September 30, 2004, Judge Fogel ruled that the plaintiffs' actions constituted fair use and that Diebold had misrepresented its copyrights under section 512(f) of the DMCA. Diebold agreed to pay $125,000 in damages and fees, but the real victory for the EFF was Fogel's succinct and forceful scolding of Diebold's legal strategy: "No reasonable copyright holder could have believed that the portions of the email archive discussing possible technical problems with Diebold's voting machines were protected by copyright. . . . The fact Diebold never actually brought suit against any alleged infringer suggests strongly that Diebold sought to use the DMCA's safe harbor provisions—which were designed to protect ISPs, not copyright holders—as a sword to suppress publication of embarrassing content rather than as a shield to protect its intellectual property."[62]

VOTING TECHNOLOGIES AND THE ELECTION OF 2004

This Diebold Debacle, as one news outlet called it, set the stage for the presidential election of 2004.[63] States started examining their machines, while reporters and political pundits around the country focused on voting machines as the most likely cause of the next electoral disaster. Some troubled implementations of newer machines sparked criticism early on: the machines in one Indiana election recorded 5,352 voters and 144,000 votes, and those in a Virginia election were subtracting votes from a candidate's total instead of adding them. Conspiracy theories and real-life anecdotes fed on each other, as commentators traded stories about machines being taken away, broken security seals in the hardware, and memory cards with voting records that mysteriously went missing.[64]

In Florida, a local election in the Palm Beach region showed that high-tech voting machines could delete the paper trails required to investigate anomalies (figure 12). Out of 10,844 votes cast in one election, one candidate won by just 12 votes, but the machines recorded 137 "undervotes." These were cases where the system failed to register a user's taps on the touch screen interface. State law required a recount, given the election's razor-thin margin, but the machine could do nothing but print the results again. There was no way to tell whether the undervotes were due to mistaken screen taps, system malfunctions, or voters' last-minute decisions to abstain.[65]

In light of these reports from Florida, the Maryland legislature contracted a firm called RABA Innovative Solution Cell (RiSC) to examine their state's electoral technologies.[66] Its director, Michael Wertheimer, was one of the country's most widely respected cryptologists. He had worked as a cryptologic mathematician at the National Security Agency from 1982 to 2003 and in 1999 had been technical director for the NSA's Signals Intelligence Directorate.[67] His report on the Maryland system was so damning that it helped push skepticism about voting machines to an all-time high. His team—essentially professional hackers—found so many vulnerabilities that they concluded that the state's election system "contains considerable security risks that can cause moderate to severe disruption in an election."[68]

Wertheimer also reported the "troubling revelation" that Diebold's code revealed a "general lack of security awareness" and that the company did not appear to have followed "widely accepted standards of software development." He agreed with the technical analyses in earlier studies and cautioned that a lot of work was necessary before the state's electoral system could be relied on to "accurately render the election" and be "worthy of voter trust." At the very least, he noted, paper records would be necessary.[69]

This report sent shockwaves across the country. Maryland had spent $55 million to purchase these machines. The entire state of Georgia and

FIGURE 12. This is a Votomatic voting machine used in Palm Beach County, Florida, during the presidential election of 2000. This photograph was uploaded to Wikimedia Commons by user Clariosophic, and I modified it only to increase its resolution. It is licensed under the Creative Commons Attribution-Share Alike 3.0 Unported License (https://creativecommons.org/licenses/by-sa/3.0/deed.en).

some of the biggest counties in California and Ohio had already adopted them. California ordered fifteen thousand of them withdrawn. Bob Urosevich, the company's president, told the *New York Times* that even though the RiSC report had confirmed "the accuracy and security of Maryland's voting procedures and our voting systems as they exist today," he saw "plenty of room for improvement and refinement."[70] There's a fabulous documentary titled *Hacking Democracy* where you can learn more about the ripple effects these problems caused.[71]

Regardless of the machines' security and reliability issues, it was too late for several states to change their plans. During the presidential election of 2004, 16.8 percent of counties used electronic voting machines such as Diebold's (up from 7.8 percent in 2000), and the percentage of total votes these machines counted rose from 12 percent in 2000 to 22.1 percent in 2004. That year, they processed 11,384,315 votes. The only vote-counting technology more popular than electronic machines in 2004 was optical scan of hard forms, which accounted for 29.6 percent of votes (26,147,292 in total).[72]

©©©

In the presidential race, George W. Bush beat John Kerry by about three million votes and twenty-six Electoral College votes. To be clear, I am not suggesting that Kerry would have won if voting machines had been more reliable. But voter confidence in elections is a frail thing, and Diebold's efforts to conceal its own shortcomings helped erode this confidence. The company seemed to cover up technological flaws by censoring its critics.

Since 2004, state and federal governments have made large investments in the reliability and security of voting machines. In 2017, Secretary of Homeland Security Jeh Johnson designated election systems a critical infrastructure despite concern in some states that this would lead to increased federal oversight. But the issue is now changing: whereas people once worried about the Diebold machines' reliability, today they are more focused election systems' vulnerability to attack.[73]

But let's get back to copyrights. Diebold's actions echo what the Church of Scientology did in the 1990s. Both organizations, when faced with the unauthorized circulation of their secret documents, turned to copyright law as a tool to attempt to maintain control over the posting, and even commenting, on these materials. The difference is that the church did so without relying on the DMCA: it was trying to carve out precedent that placed infringement liability on Internet service providers.

Diebold used a law meant to protect ISPs as a tool to enforce its own copyrights. Instead of targeting any alleged infringers, it went after the legal immunity that ISPs enjoy thanks to the DMCA. Judge Fogel ruled that this violated the law, but taking contents off the Internet is so difficult that Diebold had no other choice if it wanted to make those files disappear. I do wish, however, that the company's executives had realized earlier that they were fighting a losing battle. Perhaps this information *did* want to be free.

CHAPTER SIX

SpongeBob Goes to Court

THERE IS NO CLASS ASSIGNMENT my students enjoy more than the "Fool YouTube" project. It uses a YouTube system called ContentID, which automatically detects copyrighted material in a user's video. When you upload a video, ContentID compares its sound and video fingerprints to a database of materials that copyright holders have submitted for tracking. The system then flags potentially infringing material and gives you some options that the copyright owner has preapproved. It might require you to keep the video private or to remove the copyrighted content. Sometimes it doesn't ask you to do anything but directs all ad revenue to the copyright owner.

The "Fool YouTube" assignment asks students to see if they can trick ContentID into letting them post copyrighted material without getting flagged. They get one week to game the system and then report back to the class. The less engaged students do something simple such as playing music in the background while walking around their dorm rooms, but the ones who get excited about this come up with some fascinating ideas. Visually oriented students play with the videos' proportions, color saturation, and playback speed; the more musically inclined make covers using modified instrumental tracks and unusual accents or else modify the audio tracks themselves.

This project helps me open discussion on several copyright topics. At first, our conversations tend to be about fair use and how much a work must be modified for it to become something else. But the real discussions happen one or two lessons later, when I tell them that if ContentID flagged their work, YouTube already knows that they used copyrighted materials without authorization. (Of course, they are already reassured that what they are doing is fair use.) What other legally questionable activities might platforms know about? Some students squirm at this question, most likely because of their browsing histories. But why can we get away with things like this? What guarantees do we have that we can stay anonymous while doing it?

ContentID is part of one of my favorite stories about online copyright. In the early 2000s, after Napster and Grokster had made waves as behemoths of copyright infringement, television studios and broadcasters took aim at video-sharing platforms. Sites such as YouTube, where users could post home videos, were popular places to upload television programs. Studios and broadcasters, led by Viacom (which owned Nickelodeon and MTV and also owned SpongeBob SquarePants), launched a billion-dollar lawsuit in the hope of curbing this infringement.

This chapter tells my version of that story. Several scholars before me have written play-by-play analyses of the battle, but I'm going to focus on an unexpected outcome: neither side won, but their fight eroded user privacy in the name of copyright investigation. In addition, Google, which acquired YouTube in 2006, responded by launching the systems we know today as ContentID. This was a major accomplishment: thirty years after the Church of Scientology insisted that an automatic detection system for copyrighted works was easy to develop, Google finally launched one, and on a massive scale.

SPONGEBOB V. YOUTUBE

In February 2007, Viacom's legal team sent one hundred thousand DMCA takedown notices to YouTube, each one targeting a different video. How the company identified this allegedly infringing content is a mystery. Bloggers suspected that Viacom had simply done a general search for any term related to its properties and flagged the results as infringing, though it is also possible that the company was developing a bot that could crawl through each page. Whatever Viacom did, it was highly flawed. Reports soon circulated about Viacom targeting content it didn't own. The subjects of its complaints included a home movie shot at a barbeque restaurant, a documentarian's own film trailer, and video about karaoke in Singapore.[1]

Viacom's bold move came under immediate scholarly attack. One of the most frequently noted videos belonged to James F. Moore, a private investor and former fellow at Harvard's Berkman Center for Internet and Society. A few months earlier, Moore had posted a video of a dinner he had enjoyed at the Redbones Bar in Somerville, Massachusetts.[2] There was no Viacom content playing in the background or anything of the sort; the video simply showed a personal event that happened to unfold at a place with a name that matched Viacom's properties (likely the music of Leon Redbone).

News that Viacom had targeted this video circulated far and wide. Activists and scholars commented on the action as yet another example of corporate greed trampling on individual rights. By late February, the video had been viewed 3,243 times and had a five-star rating. It also had four comments ranging in tone from sarcastic to vitriolic:

> **moogleii**: "POS Viacom [POS could be slang for "piece of shit"]"
> **alphamone**: "indeed, die iacom die."
> **Equinoxx44**: "wow dude I such a rip from iacom!!! Oh man I see why they took it down! You are such a copycat!!
>
> "LOL corporations piss me off they see a good thing then try to capitalize on it because they cant make a greenback off it."
> **Topformfitness**: "Viacom removed my video also! I hate those fuckers . . . I made a montage of some sport contests I competed in, and a couple of scenes were televised . . . but all of myself only. I don't even think Viacom has any copywrite of this footage! How can I find out?"[3]

Media companies attacking individuals was becoming old news. The Recording Industry Association of America's individual lawsuits had become so numerous that the Electronic Frontier Foundation called them, collectively, "RIAA v. The People" and tracked their progress for many years.[4] Foundational critiques of online copyright, such as Lawrence Lessig's *Free Culture,* were written during this time. If you're interested in seeing the scope of this problem, I encourage you to read *Free Culture,* which presents dozens of stories about how corporate ownership of creative works can affect people's lives and transform what it means to do creative work.[5] These concerns were worsened by media company representatives' attitude that individual harm was mere collateral damage in their fight against piracy. As

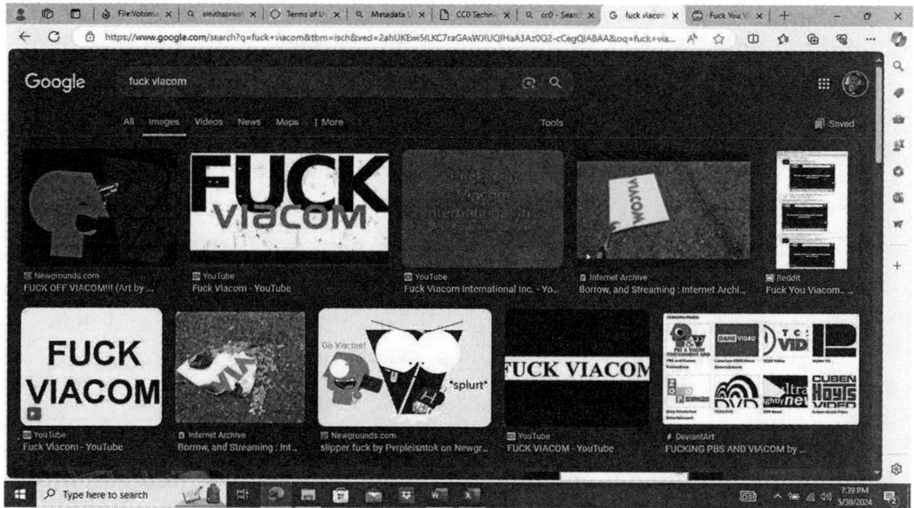

FIGURE 13. Viacom's lawsuit against YouTube generated spirited commentary among users. Folks made videos criticizing Viacom in very strong terms, accusing it of trying to shut down the web in the name of its own profits. My use of this screenshot for scholarship is protected by fair use.

RIAA spokesperson Dennis Roddy put it in 2003, "When you go fishing with a driftnet, sometimes you catch a dolphin."[6]

But there was something different about Viacom's efforts: ordinary people were being targeted for their own content. Someone like Moore knew exactly what to do with a DMCA takedown notice (file a counternotice accusing Viacom of filing a fake notice). But most Internet users were unlikely to know how to respond to this "indiscriminate spamigation" (to borrow Corey Doctorow's description in 2007).[7] Individual responses to Viacom's hundred thousand filings would not prevent this from happening again: the problem was structural, and solving it involved both developing a new system for YouTube to check for copyright infringement and broader legal changes to delineate the rights and responsibilities of all the parties involved (figure 13).

A month after filing its hundred thousand notices, Viacom and several broadcasters and studios followed up with a $1 billion lawsuit against Google. The complaint, filed at the District Court for the Southern District of New York, portrayed YouTube as a company that was almost entirely predicated on copyright infringement. According to the complaint, "YouTube has built an infringement-driven business by exploiting the popularity of Plaintiffs' copyrighted works (and the works of other copyright owners) to draw millions of users to its website. . . . There is a direct causal connection between the presence of infringing videos and YouTube's income from the additional 'eyeballs' viewing advertising on the site."[8]

The studios identified more than 150,000 unauthorized clips that had been viewed a total of 1.5 billion times. According to Viacom, there was much more copyright infringement going on because much of it remained hidden. Because users could mark videos as "private" or as exclusive for their friends, they could easily upload a television show or movie and share it privately without the copyright owner ever finding out. The studios also accused YouTube of failing to take reasonable steps to prevent piracy.[9]

One of the key examples that the studios' lawyers brought forward was a YouTube user who owned a page called "THE_RUGRATS_CHANNEL." *Rugrats,* of course, is the popular animated show about a spunky diapered baby named Tommy Pickles. The user who owned this channel was making *Rugrats* available for free. The page had a disclaimer, "Rugrats and all Rugrats related items are a copyright of Viacom." It also had a welcome paragraph explaining that they would be posting whole episodes over the coming weeks and that folks should subscribe or add them as friends because these videos might be set to private. This user's previous *Rugrats* YouTube site had been taken down over copyright infringement.[10]

It is possible that Viacom or someone with a connection to the company created this page to illustrate a point. (A few years later, Google would accuse the studios of uploading clips just to make their point.)[11] But I don't think that such subterfuge was needed, in part because the *Rugrats* videos reflect a common practice. Folks posting copyrighted material online sometimes add either a notice of the owner of the material's copyrights or else an entirely useless disclaimer along the lines of "No copyright infringement intended." Today, as back then, determined fans such as the owner of the Rugrats Channel could consider takedowns a nuisance—just something they have to deal with in order to share the content they want to share, even if it means losing an account and starting again under a new username.

In June 2007, a few weeks after the studios' filing, Google announced they would launch a suite of new technologies to flag copyrighted videos. The first component was an audio fingerprinting system that could detect content from major music studios, developed by a California company called Audible Magic.[12] This system would compare the fingerprints of user-submitted music with samples that studios had submitted to Audible. If there was a match, studios would have the option to license the content to YouTube. YouTube would then sell advertising time on the video and share the revenue with the copyright holders.[13] Google also started developing its own video fingerprinting system after concluding that there were no available commercial solutions. Its first tests would involve clips from Time Warner and Disney.

These are the systems that later became Content ID, which was soon foundational to the YouTube experience. By 2010, it had identified more than a hundred million videos and had an archive of more than three hundred thousand hours of samples provided by media companies.[14] If we think back to the Scientology and Playboy battles we saw earlier, this was the stuff of dreams. Back in the 1990s, courts released online service providers from liability on the grounds that it was impossible to implement a system of copyright infringement detection at the granular level.

Google and the studios ultimately settled out of court, after the Second Circuit Court of Appeals affirmed that the DMCA safe harbor provisions protect online platforms from liability for their users' copyright infringement. But this lawsuit is fascinating for two reasons. First, it shows how a single court opinion could either preserve the status quo of the online world or decimate it. It would be hard to imagine social media today if the court had ruled that online platforms are not shielded from liability. For better or worse, the free flow of user-generated content we know today probably would not have developed. Or it might have developed through foreign servers where DMCA enforcement would be nearly impossible, as pirated movies often circulate nowadays.

Second and more important is that the story of *Viacom v. YouTube* shows how copyright and privacy are inseparable online. The studios' effort to identify copyrighted material was blunt and limited. They did keyword searches, and perhaps they devised a way to crawl through YouTube, but their results were not complete. Users were sharing content with each other privately, and their profile pages did not necessarily have enough information for the studios to go after them individually. This meant that the studios needed information that only Google had: users' IP addresses and, most important, lists of users who had watched a given video. The case directly pitted users' right to privacy against the studios' copyrights.

GOOGLE'S SECRETS

Personal privacy was never the target of this lawsuit, but it was a casualty. A crucial turning point occurred in the summer of 2008, when the secrecy around so much of Google's inner workings came into conflict with the discovery processes that lawyers use to create evidence. To understand what happened, we need to think about a different form of intellectual property protection: the trade secret. Trade secret laws in the United States used to vary from state to state, although starting in 2020, legislatures have been working toward national uniformity by enacting a law called the Uniform Trade Secrets Act. Generally, though, information qualifies as a trade secret of the organization if it is commercially valuable and known to only a few people and if the organization takes reasonable steps, such as confidentiality agreements, to keep it secret.

We have all run into trade secrets. Think, for example, about the code that runs Google's search engine or the eleven different herbs and spices in KFC chicken. Trade secrets can be as simple as the location of something: a family business is the sole supplier of the naturally sourced mud that Major League Baseball uses to make baseballs easier to grip. The location of this mud is a trade secret.[15] Trade secrets can also be controversial. Recently, high-tech firms responded to legal threats by arguing that the diversity statistics of their workforces are trade secrets. This is ironic: Silicon Valley firms are covering up their lack of diversity by implying that knowledge about a workforce's diversity can create a competitive advantage.[16]

The studios wanted to get ahold of one of the most important trade secrets in Silicon Valley—Google search engine source code. They thought that if they could analyze this code, they would be able to determine, once and for all, whether Google (and YouTube) tended to favor copyright-infringing content over legal user-generated content. Their lawyers seem to have imagined that expert programmers would find some snippet of code or design a test that would show that the results users got when they typed something into YouTube's search bar tended to show infringing content first. They were likely incensed when Google announced it had developed video ID software. Its code was also a trade secret, and the studios wanted it. If any code was likely to show that Google's software could identify copyrighted materials, this was it.[17]

Google was also holding on to materials it considered confidential even if, strictly speaking, they didn't qualify as trade secrets. The studios were particularly interested in databases called "User" and "Mono." These held information about each video, including title, keywords, public comments about it, whether it had been flagged as inappropriate (for obscenity and copyright infringement, among other things), whether any administrative actions had been taken, and whether the user who posted it was terminated. The studios also wanted the schemas for Google's advertising and video databases. A schema is a sort of table of contents for a database—a listing that gives a sense of what the database has without showing all the data.

But perhaps the most personally sensitive information the studios wanted was something called the logging database. This was not a trade secret, but it was highly confidential because it contained plenty of personal information. Each time a video was watched, YouTube recorded the username of the person who watched it, the time when they watched it, the IP address of any devices used to watch it, and the identifying information about the video itself. It kept this information in a database on its hard drives. Using that database, one could reconstruct the viewing history of any video at any time. The studios wanted access to this database so that they could compare the attractiveness of copyright-infringing videos versus noninfringing content. They also wanted information on which private videos individual users had posted in order to compare these lists with the same users' entries in the logging database.

Google and the studios agreed to a protective order that would apply to all sensitive information that each of them provided as evidence. The twenty-page order included some extreme measures, such as the stipulations that anything marked "HIGHLY CONFIDENTIAL" would be delivered in a sealed envelope to the court and that only a properly authorized agent of the court would be allowed to open it. The details of how the parties agreed on this order remain confidential, but the agreement makes it clear that Google lobbied very aggressively to protect confidential source code and seems to have paid far less attention to personal information.

User information is mentioned only once, as "personally identifying information concerning users (which shall not be construed to include any information a user posts or uploads for public viewing on any website)."[18] It is an item in a long list that also includes business strategies, valuations of investments, nonpublic applications, and revenue share and pricing information. The order stipulates that any of these things "may be designated

HIGHLY CONFIDENTIAL by the producing party or a party who is the original source of the document or information."[19]

Source code, by contrast, gets nearly three full pages. The order allows a party to designate any instances of it as "RESTRICTED CONFIDENTIAL-SOURCE CODE." All source code would be provided on a portable hard drive, and it could be encrypted if the party providing it wished it to be. Access would be restricted to "Source Code Custodians" and experts retained to analyze it. It could be viewed only on non-networked computers in secure locked areas of the provider's offices. Outside these offices, the Source Code Custodians would be able to view the code only at sites where depositions related to the code were taking place, the court, or intermediary locations (such as a hotel before testifying), as long as the custodians took the necessary precautions.[20]

The district judge presiding over the case was Louis L. Stanton. The studios told him that Google was not cooperating with their investigation and demanded that Google hand over its secrets. The details of their battle over this evidence remains confidential, but it is clear that Google fought very hard to protect its source code and paid little attention to anything else. The only things Stanton granted the studios were the videos that had been removed from YouTube, the schema for Google's advertising database, and the logging database. The code was protected, but user data was not.[21]

DISMISSING PRIVACY

The relations between commerce and surveillance online are a major field of study. A business scholar named Shane Greenstein, for instance, wrote a book titled *How the Internet Became Commercial* in which he shows how the transformation of the web into a commercial tool created a race to ensure that users engaged directly with the ads they saw.[22] Another scholar, Shoshana Zuboff, documents how the data-intensive techniques that companies have used to achieve commercial engagement are so well established and pervasive that they form the core of a whole new economic system she calls "surveillance capitalism."[23]

The early 2010s, when Stanton considered Viacom's complaints, were a transitional period that connected the phenomena Greenstein described with those that interested Zuboff. It was clear that corporations were developing an unprecedented program of mass surveillance. Several reports had emerged about how Google, Microsoft, Yahoo!, and several other digital giants created profiles of individuals based on their browsing habits. Websites would leave behind small computer files, called "cookies," on users' computers—and these files would allow companies to infer information about the user and predict some of their decisions. The goal was to show the user the ads most likely to make them click on a link or purchase something.

I was in graduate school at this time, and I remember feeling very worried because corporations seemed to be headed toward a private version of a program the US Defense Department once called "Total Information Awareness." Launched in May 2003, TIA was a massive system that would predict terrorist threats based on data collected about everyone in the United States, including government records, travel history, purchasing habits, personal connections, and medical and financial history. Congress defunded the project that same year after an onslaught of media criticism, but its legacy lived on in the National Security Agency.

Like many other scholars in the 2010s, I worried that digital giants were getting to know far too much about us. But the problem was not just that they were gathering this information. It was that we had no way of knowing what they knew about us or how that information moved within and between firms. As the old adage about online services goes, "If the product is free, you are the product."

In this context, Google's logging database played an important symbolic role. Every YouTube user has a viewing history that details everything they have watched; the logging database cut across all these viewing histories to report all the users who had watched a specific video. The immediate concern was that if this were given to Viacom, it would have access to a list of users who had infringed on its copyrights. But the bigger issue was that ordering Google to release its logging database could create a precedent for demanding that online service providers identify the users who had engaged with any legally contentious content. If we could no longer view certain content with at least some assurance that our actions would not be available for scrutiny by others, our browsing habits might change forever.

This concern makes Google's defense against the studios' demand for the logging database very disappointing. To me, it was stunningly weak. For instance, Google's lawyers insisted that it would be too "extensive and time-consuming" to provide the information, because it would need to screen the contents "for privileged and work product material."[24] Google's lawyers also pointed to the Video Privacy Protection Act of 1988, which prohibited video cassette rental companies from disclosing what materials subscribers request or obtain. It cited a recent case involving Amazon, in which a federal court in Wisconsin rejected a motion to disclose users' reading habits to the government on First Amendment grounds. Judge Stanton ruled these cases inapplicable, most likely because one involved video cassettes, not online videos, and the other involved disclosure to the government. He ordered Google to produce "all data from the logging database" concerning all YouTube videos. He wrote, without any legal analysis, that "defendants cite no authority barring them from disclosing such information in civil discovery proceedings, and their privacy concerns are speculative."[25]

Stanton dismissed Google's privacy concerns as speculative and granted the studios' motion to acquire the logging database. By Google's own estimate, there were about twelve terabytes of logging data, an amount easily transported in three or four commercially available hard drives. The studios could now access any video's viewing history and identify the users who had committed some form of copyright infringement. There was no requirement that they inform users or provide any opportunity to contest the information the studios would obtain about them.

This small decision had immense consequences, in that it would have exposed the viewing history of every YouTube user. Stanton's decision seemed to indicate that user privacy was simply not a concern in his eyes. It also leads me to speculate that Google simply did not defend user privacy as determinedly as it defended other aspects of its operation that it deemed more sensitive. If Google's attorneys had dug deeper into the Video Privacy Protection Act, they could have argued that the act's protection of video cassette owners also extended to online videos. As the Electronic Frontier Foundation was quick to point out, the act applied to "prerecorded video cassette tapes or similar audio visual materials."[26]

Perhaps this is a moment where a stronger metaphor could have made a big difference. Recall how the Scientologists' battle against Netcom in the 1990s boiled down to a battle of comparisons: What is the online flow of information like? Is it a river, or a bookstore? The court had absolved online service providers of liability partly because it adopted the notion that computer networks were like a swap meet. It made no sense to shut down the whole swap meet just because some users brought in unauthorized materials. Perhaps if Google had pressed harder on the metaphor of

YouTube as a video rental service, Stanton would have been inclined to apply the Video Privacy Protection Act to this case.

Kurt Opshal, an attorney at the Electronic Frontier Foundation, explained this connection in July 2008. He wrote: "The act refers to 'prerecorded video cassette tapes or *similar audio visual materials.*' A YouTube video may not be a videotape, but certainly qualifies as audio visual material. Thus, YouTube is a 'video tape service provider' under the act, because it is 'engaged in the business [of] delivery of . . . audio visual materials.' The VPPA protects 'personally identifiable information,' which is defined to include 'information which identifies a person as having requested or obtained specific video materials or services.' This is exactly what is in the Logging database."[27]

This metaphor is not perfect. For one, YouTube users can watch videos for free, so their relationship with the company is different from the relationship between a renter and a video store. The metaphor thins out further when we consider that YouTube users can submit content to the platform, whereas renters could not provide a video store with new tapes. But that's the point of metaphors—they allow us to grasp a concept for which we have no preexisting understanding. Even if the video rental metaphor had been unavailable, something else could have worked: a video swap meet, perhaps?

If Google's lawyers had pressed harder on the Video Privacy Protection Act, they could have at least pointed out that the spirit and scope of the act protected sensitive personal information about users' viewing histories. The act prohibited courts from ordering the disclosure of personally identifiable information regarding video use unless there was no other information that could meet a need, consumers were given reasonable notice, and consumers were given the opportunity to contest the claim. By bypassing this reasoning altogether, Stanton essentially stripped YouTube users of privacy in their audiovisual consumption that they could have reasonably expected to have.

The damage was done. YouTube users' viewing habits were made available for the studios to examine. To avoid further public outcry, Viacom stated publicly that it would not go after individual users. The studios also agreed to have Google anonymize all user data (though, as scholars of data know well, it is very difficult to fully anonymize personal information). But importantly, these two protective measures were not legally required. They were additional steps taken largely because advocacy groups were watching the situation and holding Google and the studios accountable. Viacom's lawyers even told the EFF that they would notify it in advance if they required any changes to the protective order.

WHO KNEW?

Judge Stanton's dismissal of user privacy did not help anyone. The user data, once anonymized, became useless to the studios, and I found no significant discussions of it in later court documents. This whole battle over Google's secrets was an effort to prove that YouTube knew and incentivized copyright infringement on its platform. Recall that there are several kinds of copyright infringement. One is direct infringement, such as when a person posts an unauthorized movie to YouTube, but there are also contributory infringement, which occurs when someone knows the infringement is happening and contributes to it, and vicarious infringement, when someone could control the infringement but is benefitting commercially from it. Napster had been found guilty of the latter two.

Stanton quickly dismantled the possibility of holding YouTube accountable for any of these. In 2010, he ruled that Google was protected by the safe harbor provisions of the DMCA. His opinion acknowledges that YouTube had general knowledge about infringement happening on its platform, but its employees did not know which videos had been uploaded without the copyright holders' consent. This aspect of his opinion echoes the Scientology case very closely: Stanton explained that requiring a video-sharing site to monitor every video for copyright infringement was untenable and "would contravene the structure and operation of the DMCA."[28] This was by far the most vulnerable part of Stanton's opinion; the studios would simply need to show on appeal that YouTube employees knew of specific instances of copyright infringement.

Stanton also rejected the studios' comparisons between YouTube and peer-to-peer services such as Napster and Grokster. A smoking gun he used to do this was an email that Viacom's general counsel had sent in 2006, stating that "the difference between YouTube's behavior and Grokster's is staggering."[29] YouTube had not distributed its platform with the sole aim of encouraging copyright infringement. Instead, from the very beginning, the platform had been designed and marketed to share personal videos. That, perhaps, is the point of the "You" in "YouTube." To echo the Betamax lawsuits from the 1980s, YouTube had substantial noninfringing uses.

Viacom filed a quick appeal, and the case entered a long round of hearings that culminated back at Stanton's court. The studios came armed with documents which, they claimed, showed that YouTube staff knew about the rampant copyright infringement happening in their platform. They had been able to obtain these messages because user trends came under consideration when Google's board of directors was thinking about acquiring YouTube. According to the studios' attorneys, in March 2006, one of YouTube's founders (Steve Chen, Chad Hurley, or Jawed Karim) had sent a message to the board acknowledging "blatantly illegal" copyright infringement that included Viacom content. These emails are unfortunately unavailable to me for verification.[30] But they were enough for the US Court of Appeals for the Second Circuit to send the case back to Stanton for reconsideration.

Back in Stanton's court, Google won yet again. General knowledge of infringement was different from the granular level of detail that systems such as Content ID (announced and launched after the lawsuit started) would be able to provide. Once again, as in the Scientology case, the volume of material hosted on YouTube made it impossible to check for infringement one post at a time. Even Viacom and its allies acknowledged this, having told the court that "it has now become clear that neither side possesses the kind of evidence that would allow a clip-by-clip assessment of actual knowledge."[31]

The studios prepared another appeal, but the case ended in 2014 with an out-of-court settlement. Viacom would move on to a different major legal battle, a class action suit from parents who claimed that the company was violating their children's privacy. And Google would continue to grow, carving out increasingly robust protections for online platforms to avoid liability for copyright infringement. But their stories, intersecting in this lawsuit, eroded user privacy just as viewing habits and user trends were becoming social media's bread and butter.

©©©

My "Fool YouTube" course assignment was born about six years after this case. I started thinking seriously about Google's copyright policies and ContentID during the covid-19 pandemic. During my performing heyday as a classical baritone, many years ago, I had been recitalist and soloist at places ranging from small New England churches to Costa Rica's biggest theaters. In the years after *Viacom v. YouTube,* I started exploring more contemporary singing styles using backing tracks on YouTube. I knew how the ContentID system worked, but I have always tried to keep my artistic practice away from my scholarly one, and I had no interest in thinking about copyright while singing unless my playlist of Josh Groban backing tracks was compromised.

When the pandemic hit, my vocal coach encouraged me to share my music with local community groups through YouTube. I asked a friend to rip YouTube backing tracks for Andrea Bocelli's songs "Fall on Me" and "Canto della Terra." Then my coach quarantined long enough to be able to visit our home and set up a recording studio for a day. It was a beautiful day—one of my happiest during the pandemic, as far as I can remember. I set the video settings to "public" for a few days so that the members of a local spiritual community could listen to the songs, and a few days later I set them to "unlisted" so that only that community and some of my closest friends could listen to them.

When I visit my YouTube Studio today, I can inquire into each of my videos' copyright situation. The system informs me that one of my videos uses the melody from "Fall on Me." The owners of the melody are listed as "CD Baby Sync Publishing, LatinAutor, LatinAutorPerf, União Brasileira de Compositores, UMPG Publishing, UNIAO BRASILEIRA DE EDITORAS DE MUSICA—UBEM, UMPI, LatinAutor—PeerMusic, LatinAutor—UMPG." This listing of owners is fascinating because it traces the history of the backing track. My friend didn't rip the music from an official video from UMPG (Universal Music Publishing Group). Instead, they ripped a video featuring a special piano arrangement of a song. The other "owners" include "CD Baby Sync Publishing" (which licenses music for synchronization with videos) and firms that license Latin performances and musical arrangements.

All these owners could have prohibited me from making this video, and they could still change their mind. The "Copyright Details" page notes: "The ContentID claim on your video doesn't affect your channel. This is not a copyright strike" and states that the claim has "No impact" on my video unless I try to monetize it. This is where YouTube tells me the kind of rights that the owners of the melody are willing to grant me. (I should note that some folks making backing tracks use audio watermarks such as the ones I discussed earlier to catch folks who rip them without permission.) UMPG and its peers could have easily prohibited the melody from being used at all, or to restrict its use only to "unlisted" videos. Luckily for me, I was able to sing my song, and I walked away with a new idea of how to teach.

On January 5, 2021, a San Francisco research laboratory called OpenAI announced a new digital image generator called DALL-E. It's a fantastic name—a blend between Spanish surrealist painter Salvador Dalí and Wall-E, the adorable Disney-Pixar robot who made my father feel an emotional connection to an animated film for the first time in his life.

DALL-E was the first of several text-based image generators. It works quite intuitively: you write a description of what you want into a text box in your Internet browser, and a few seconds later the system gives you four images showing what you wanted. My experiences with DALL-E brought me to one of the biggest unsolved puzzles about copyright on the Internet: What do we do about systems that learn from the things we post online?

When I first saw DALL-E, in the spring of 2023, my first instinct was to play with my own memories. I was sitting in a hotel lobby missing my dogs, so I asked DALL-E to paint one of them. The images in figure 14 are a photograph of our dog Kupo and what DALL-E gave me in response to a verbal description. I was stunned. The image on the bottom is not a real dog. It's a representation of one that DALL-E generated, but I am sure I have seen Kupo sleeping in the exact same pose.

Playing with DALL-E some more, I started to wonder about bias and representation, and whether DALL-E would show the racial biases that Safiya Noble analyzed in *Algorithms of Oppression.*[1] Noble described how

Google Images' results page exhibited racial biases by showing, for example, mugshots when asked to deliver images of Black men. I was relieved to see that prompts such as "police mugshot of a man" did not deliver primarily Black men. But the relief vanished when the prompt "the most beautiful person in the world" delivered the following images: three white women with blue eyes and a light-skinned baby.

I was using DALL-E 2, which OpenAI will sunset in 2025. Its replacement, DALL-E 3, which is integrated into ChatGPT, responds to the prompt "draw the ugliest person in the world" with the following: "Drawing someone to be intentionally ugly would be disrespectful and offensive. As an AI developed by OpenAI, I strive to promote respect and positivity in all interactions. If you have a different request or need help with something else, feel free to ask!" To record the existence of this bias in DALL-E 2, I created the images in figure 15 on May 30, 2024.

From a technical perspective, this result wasn't surprising. DALL-E uses a technology called CLIP (contrasting language image pre-training), a kind of image recognition system that helps DALL-E rank the images it creates and offer the four best matches. Unlike earlier recognition systems, which used labeled datasets of images, CLIP uses online images and their captions: it learns what a chair is, for example, by processing images whose caption describes a chair.

From Noble's work, we might think that the biases embedded into online captions would filter into CLIP's ranking system. This is not to say that CLIP *creates* bias, only that it reflects ideas of beauty online by processing images with captions describing a "beautiful person." These ideas are gendered and racialized, and they also carry expectations, such as that a beautiful person will have clear skin and long hair and pose for a selfie. But what's more important is that these concepts of beauty are not special as far as DALL-E and CLIP are concerned: the systems are designed to identify and perpetuate visual-textual patterns of all kinds.

FIGURE 14. The top image shows our dog Kupo, who often sleeps in this position in our living room. I took this photograph. The bottom image is an AI rendering I made using DALL-E 2 with the prompt "a shih-tzu and lhasa-apso mix dog, mostly tan color, sleeping peacefully on a fuzzy dog mattress that is white and gray." The Copyright Office does not recognize AI-generated works, so I did not need permission to include the second photograph.

FIGURE 15. This image documents how DALL-E 2 could respond to prompts in racialized ways. It generated the top series in response to the prompt "most beautiful person in the world" and the bottom series in response to "ugliest person in the world." These images were created by AI, so I do not need copyright permission to use them here.

Back in spring 2023, I kept asking DALL-E for images of the most beautiful person in the world, and by the third attempt I started getting images of people of color: two people who appear to have Latin American descent. It took six attempts before an overweight person and one with East Asian features showed up (a bearded man and a baby, respectively). The converse is also true: a search for "the ugliest person in the world" yielded two white men with receding hairlines (one obese, the other with misaligned teeth), a cartoon of a bald white man with missing teeth, and an overweight Black woman yelling at the camera. This pattern continued for several more attempts. White skin, receding hairlines, bad teeth, and acne were by far the most common responses. A Latino man and another Black woman appeared in my fifth attempt—both overweight, with unusually shaped teeth, scowling.

This perpetuation of stereotypes is not free from human intervention. In March 2023, OpenAI executives admitted to implementing safeguards that prevent their text-generating system, ChatGPT, from generating offensive responses to users' prompts. By some accounts, ChatGPT would not generate a positive poem about Donald Trump on the grounds that it was "partisan, biased or political content," but it easily generated one about Joe

Biden.[2] Similar filters are in place at DALL-E: it will not generate images depicting murders, sexual encounters, specific political figures, or a long collection of categories listed in its content policy.[3] It is worth noting, however, that a request for "a cartoon depicting a president" led me to four pictures of Black men in suits with short hair like Obama's and one that looks like a cross between Donald Trump and Vladimir Putin. So, according to my searches, its training database appears to be politicized even if users cannot request politicized content.

My experience is entirely unsurprising given recent scholarship in algorithm and Internet studies. What interests me most about this play date with DALL-E is that every image I saw was based on other people's creative works. DALL-E, like all other artificial intelligence text or image generators, is trained by feeding on billions of texts and images. When it creates a new image, it is mimicking, blending, and reshuffling all these previous works. Then CLIP ranks DALL-E's output based on its analysis of text-image pairings online. Other people's creative works had to be taken from the Internet, uploaded to a database, processed by these systems, and used as the basis for creating and classifying new works.

This process can yield images that bear an artist's (or a corporation's) distinctive style. The prompt "paintings of people in the style of pablo picasso," for example, generated a mix of cubist-looking portraits. The prompt "mickey mouse" led to a collection of 3-D renderings of Mickey that brought together features from his different styles across the years. What are the copyright implications of this process? If Picasso were alive today and wanted to stop DALL-E from generating images in his style, what could he do? And could OpenAI argue that it owns the copyrights over DALL-E's pictures of Mickey Mouse?

The pictures that DALL-E generated are all original at least in one sense: they are not direct copies of anyone's work. If DALL-E had spat out a direct copy of *Les Demoiselles d'Avignon* or a screenshot from *Fantasia*, then certainly we would have a case for copyright infringement. (It's worth noting, though, that this case would be much weaker if DALL-E *generated* such an image from scratch. Copyright law does allow for independent creation of the same work.)

But the images DALL-E showed me were *in the style* of Picasso's and Disney's works. Copyright law does not necessarily protect styles of work: no one can own cubism, although they can own their own creative expressions in cubist form. Disney does own copyrights over several versions of Mickey (the first one, from the film *Steamboat Willie,* expired in 2024), but I have never seen any Mickeys like the ones DALL-E gave me. Who owns these images? And is it copyright infringement for DALL-E to use people's works to generate new works without their consent?

©©©

To think about this question, it is helpful to consider a monkey named Ella and known in legal documents as Naruto. In what is perhaps my favorite legal case of all time, a US federal court ruled in 2018 that a monkey did not own the copyrights to a selfie it took in the forest. Seven years earlier, a British photographer named David Slater had gone to an Indonesian forest to capture images of a critically endangered species of monkey called the Celebes crested macaque. He let the monkeys discover and play with his cameras while they were turned on, and inevitably one of them photographed herself (figure 16).

As part of his marketing strategy, Slater claimed that the monkeys had taken the pictures, even though he was very involved in the process: he placed a camera on a tripod and held on to it while the monkeys moved the camera around. He had also installed remote shutter triggers that the monkeys would inadvertently hit and had put the camera on autofocus. These were *his* pictures, he thought, because he had set up the environment where the monkeys' actions would generate portraits according to *his* vision. (Had the monkeys taken a camera from his backpack and started playing with it on their own, as some rumors eventually stated, then Slater might have had no justification to claim the pictures as his work.)

Slater's photography business relied on licensing revenue, which is exactly the kind of thing copyright is meant to protect. But when he filed takedown notices after an image was posted without his consent, online venues such as the Wikimedia Foundation (which runs Wikipedia) refused to comply on the grounds that copyrights are for humans: since a monkey took the picture, Slater had no copyright claim over it. The foundation seemed to relish this fact; at Wikimania 2014 (the annual conference celebrating the Wikimedia Foundation's projects) attendees could even take selfies with printouts of the monkey's selfie.[4]

Slater's copyright battles did not end there. In 2015 the environmental activist group PETA (People for Ethical Treatment of Animals) filed a lawsuit arguing that one of the monkeys, whom PETA named Naruto, owned the

FIGURE 16. Photographer David Slater named this monkey Ella, but court documents for a lawsuit over the copyrights for this image—did she take it or not?—renamed her Naruto. The rights over this image are handled by Caters Media Group, a UK firm. I purchased permission to include it in this book and was asked to include this credit line: Caters Media Group.

copyright to her selfies. The case was called *Naruto v. Slater.* An appeals court eventually ruled that "the monkey lacked statutory standing because the Copyright Act does not expressly authorize animals to file copyright infringement suits."[5] It did not say that animals couldn't bring suit—their inability to do so was reaffirmed by the unusual case of *Cetacean Community v. Bush* (2004)—rather, it found that nonhuman animals could not count as authors under copyright law.

But the court was even more specific: it ruled that copyright law is meant *only* for humans. This conclusion was grounded on the Copyright Act's implicit descriptions of what counts as an author. The act used terms such as "children," "grandchildren," and "widow" and described situations such as an author's widow or widower owning one half of the author's interest. According to the court, all these terms "imply humanity and necessarily exclude animals that do not marry and do not have heirs entitled to property by law."[6] Slater and PETA ultimately settled the case, and Slater's legal expenses brought him to the brink of bankruptcy.

This mode of reasoning was not surprising. When deciding *Roe v. Wade* in 1973, the Supreme Court read the US Constitution to conclude that its definition of "person" includes being born. The justices affirmed this close reading in 1989, explaining that the words of a statute "must be read in their context and with a view to their place in the overall statutory scheme."[7] In other words, courts are required to understand a word in the context of its use in legislation or the Constitution.

The restrictive definition of authorship is not surprising. More than a century ago, the Supreme Court upheld the copyrightability of photographs because they were "representatives of original intellectual conceptions of the author." In the process, it defined an author as "he to whom anything owes its origin; originator; maker; one who completes a work of science and literature." The author was a *he*—by which they meant a person (whom they assumed to be a man by default). But photographs were copyrightable only as long as they were not "merely mechanical" processes: if they had "no place for novelty, invention or originality," then "a copyright is no protection."[8]

Back to the monkey selfie case and our playdate with DALL-E: since settling the case with PETA, Slater now uses his website to sell assorted goods with Naruto's picture printed on them. I have a Naruto postcard and gave a print to a friend. But what's most important about the Naruto case is that it affirmed a principle that is at the core of AI copyright today—only humans can own a copyright. The Copyright Office has been interpreting this to mean that for a work to be copyrightable, *all* authorship must be human. Its *Compendium* of rules lists "a photograph taken by a monkey" alongside "a song naming the Holy Spirit as the author" as examples of works that are ineligible for copyright protection. It also explains that the Copyright Office "will refuse to register a claim if it determines that a human being did not create a work."[9]

This became apparent on February 21, 2023, when the Copyright Office rescinded registration certificates for artwork that a comic book artist named Kris Kashtanova had created with the help of Midjourney, an AI art generator (figure 17).[10] Earlier efforts to register AI generated art had failed because the humans who "made" them stated that they had prompted a computer to make a picture.[11] Kashtanova challenged this by submitting extensive documentation of how she used Midjourney as an aid in her artistic process: Midjourney would generate an image, which Kashtanova would then resubmit to the AI with additional text prompts until she got exactly the image she wanted. This didn't matter to the Copyright Office. If an AI system generated an image, no copyright can be granted—regardless of how much an artist reworked it.

This leads us to a major legal problem: What kind of protection is available for AI-generated images? Most nations in the World Intellectual Property Organization, including Australia, Spain, Germany, and the United States, generally require a human creator for copyright protection. However, the United Kingdom, India, Ireland, and Hong Kong assign ownership of an AI-generated work to the AI's programmer.[12] In the United Kingdom, for example, the Copyrights, Designs and Patents Act states that "in the case of a literary, dramatic, musical or artistic work which is computer-generated, the author shall be taken to be the person by whom the arrangements necessary for the creation of the work are undertaken."[13] This generates great ambiguity: Who counts as the person who made these arrangements?

In the United States, courts, scholars, and artists are still working this out. Some folks argue that the makers of the AI system should hold the copyrights, but that is not how copyright works in the United States: it is designed to incentivize creators, not the makers of the technical systems that allow them to create. As one artist asked me during casual conversation at a conference, would you want the maker of your paintbrush to have rights over your painting? But even if copyright *could* work like this, what does it mean to *make* an AI system? These systems are written by multiple people using algorithms that other people wrote—and they are trained on databases compiled by yet other groups of people.

FIGURE 17. This is the cover of a beautiful comic book created by artist Kris Kashtanova using Midjourney. This image is in the public domain. Read the comic at www.kris.art/portfolio-2/project-one-ephnc-jamy8.

Other folks argue that AI-generated art should be treated as work-for-hire content. When a company hires an artist to create something under a work-for-hire agreement, the rights belong to the company. In this logic, the AI is an artist, and the company that launches it is treated as the one that hired the artist. Legally speaking, this appears to be a more reasonable approach, if only because it doesn't require us to figure out who the creator of these works should be and bypasses the copyright-eligibility issues raised by cases such as *Naruto*. Whether this can work remains an open problem— one that will probably require congressional intervention.

Even if we could magically resolve the matter of who owns AI art, we are left with a legal problem: Should artists have the right to exclude their own work from the training databases these AI systems use? Let's think about Picasso again. The only reason DALL-E was able to generate portraits of people in his style is that it had processed a bounty of images labeled Pablo Picasso. There is no way to remove all those images from the Internet and prevent more images labeled "Pablo Picasso" from being posted every day. Picasso could be removed from DALL-E only if its makers explicitly filtered out images with the right collection of Picasso-related keywords. But could Picasso's estate legally force OpenAI to do this?

The stakes are high for many artists who depend on their own distinctive styles to sell their art and market their merchandise. Consider, for example, one of my favorite independent art brands, Kimchi Kawaii, at www. kimchikawaii.com. Founded in 2008 by a South Korean immigrant now named Holly Guenther, Kimchi Kawaii sells products by "a Korean girl doing Japanese 'kawaii' (cute) inspired art."[14] Gunther is a regular at comic book conventions; I visit her booth at conventions' artist alleys every time my husband and I attend together. She is known for color-blocked images of an adorable tiger in traditional South Korean garb and a line of illustrations called Purrista Pawfee: cute images depicting drinks and baked goods as cats, some with punning names such as Catpuccino and Mewcaron. She sells her art as prints and on an assortment of products, from pins to coffee mugs.

Could Guenther do anything to force AI art generators to exclude her work from their training databases? If she simply stopped tagging images as "Kimchi Kawaii," the exposure her art received on social media would plummet. But if she keeps tagging her images with her brand name and her popularity explodes, DALL-E and its counterparts might start drawing drink-cats in the style of Kimchi Kawaii. Luckily for her, a DALL-E request for "a kawaii image depicting a Starbucks drink as a cat" does not yet yield a copy of her style. And requesting images using the words "kimchi kawaii" gener-ated cute cartoons of kimchi jars. It appears that DALL-E has not yet been trained in the subtleties of Kimchi Kawaii art as it has been with Picasso's, but that could easily change.

Once artists achieve some popularity, it becomes incredibly easy to generate art in their style. One page in the popular site Metaverse Post was titled "Midjourney and Dall-E Artist Styles Dump with Examples: 130 Famous AI Painting Techniques."[15] The page depicted art-making as a writing exercise, with such statements as: "It's difficult to employ proper text-to-art suggestions in an artistic manner."

According to the website, folks who wanted to write their own prompts would simply follow this process:

Step-by-step guide:

1. Select an artistic style, such as Kazimir Malevich's
2. Select between portrait and landscape orientation
3. Include more information and details like: ultra realistic beautiful skinny man with katana sword
4. Select aspect ratio is 9:16.
5. Your first Artist Style Prompt is ready.[16]

This would generate a statement such as "Portrait of ultra-realistic beautiful skinny man with katana sword by Kazimir Malevich—ar 16:9." If this seemed too complicated, Malevich fans could simply go to any of the six free text generators designed specifically to turn their descriptions into optimal prompts. The page also listed 130 styles of art, each featuring two to four artists whose work could help a user generate a prompt in the corresponding style. These artists' styles were entirely recognizable and easy to duplicate using AI art generators.

In the past couple of years, artists have banded together online to oppose AI art, and in January 2023 several of them filed a class action lawsuit against the AI art generators Midjourney, Stable Diffusion, and DreamUp.[17] Like DALL-E (which was not listed as a defendant), these generators use a nonprofit database that indexes more than five billion images found online. The artists' complaint has three components: they did not agree to have their work included in the database; they were not compensated even though the companies that made the AI charged for using their systems; and they were not credited.

The case was dismissed, but one thing will surely come into play in future efforts: the notion, popularized by Google in the early 2000s, that the inclusion of a copyrighted work in a database transforms it from an artistic work into a functional one—a data point in a database. This notion can seem counterintuitive, and there is plenty of evidence to argue that it is explicitly designed to advance Google's interests and nothing else. But without it, we could not have image-based searches—and we would therefore not have AI art generators such as DALL-E.

This final section of the book explores how this notion became a central pillar of copyright law. It opens not in twenty-first-century Silicon Valley, as we might expect, but in 1980s Miami, where a notoriously profane rap group called 2 Live Crew won major victories for both free speech and fair use exceptions to copyright infringement. In the final two chapters I'll show how these victories allowed giant companies (especially Google) to advance a legal theory of transformative uses that benefits online service providers instead of the content creators whom copyright was meant to protect.

CHAPTER SEVEN

Rappers to the Rescue

IN THE LATE 1980S, FOUR MIAMI performers—DJ Mr. Mixx, Brother Marquis, Fresh Kid Ice, and their leader, Luke Campbell—became famous for their sexually explicit lyrics. They were part of a group called 2 Live Crew, and within a few years they were among the most successful artists in the Miami bass style of hip hop music, thanks in no small part to a high-profile legal battle that ultimately took them to the Supreme Court. The battle revolved around two albums that the group released almost simultaneously: an explicit one titled *As Nasty as They Wanna Be* and a "clean" version called *As Clean as They Wanna Be.*

These albums made 2 Live Crew the most important musical group in the legal history of censorship and copyright. In the 1980s, sexually explicit lyrics in popular music were a frequent target of politicians' rage, and 2 Live Crew's work was no exception. With songs such as "The Fuck Shop" and "Dick Almighty," *As Nasty as They Wanna Be* triggered Florida politicians' wrath. In 1990, the sheriff of Broward County, Nicholas Navarro, even threatened to arrest the owners of music stores that sold the album. The group sued the county and won a judgment that was both a relief and free advertisement: a judge ruled that the album "is hereby DECLARED obscene" and that the sheriff's actions were an unconstitutional restraint of free speech.[1]

But the reason I am interested in 2 Live Crew concerns one of the songs in their other album. When Luke was writing songs for *As Clean as*

They Wanna Be, he decided to parody the classic song "Oh, Pretty Woman," written by Roy Orbison and William Dees in 1964, which famously repeats the phrase "Pretty woman" several times. In the middle of the song, for example, Orbison sings: "Pretty woman, walking down the street / Pretty woman, the kind I'd like to meet." I can't quote more from the song to avoid legal trouble (even despite fair use protections), but you can Google the song title and find several sites with the lyrics. This is yet another contradiction in our copyright system.

Luke's parody infused the classic with 2 Live Crew's irreverent and misogynistic style. In his hands, the song comments on the looks and attitudes of several women, including a "pretty woman," a "Bald-headed woman," and a "Two-timing woman." The second stanza (once again truncated), goes: "[Big hairy woman] You know, I bet it's tough / [Big hairy woman] All that hair, it ain't legi-i-it."

Lyrics like these wouldn't get the group in trouble with censor-hungry sheriffs. There was also a well-established procedure to secure the rights needed to use another artist's work: the mechanical license, which grants permission to reproduce and distribute a musical composition. The Copyright Office sets the minimum royalty rate for these licenses, increasing them periodically to adjust for inflation. In the late 1980s, the minimum required rate was about 4.25 cents per copy of the song distributed.[2]

Following procedure, Linda Fine, 2 Live Crew's manager, contacted Gerald Teifer, a music executive who handled licensing at Acuff Rose Music, the label that owned the rights to Orbison's song. In a letter, she told him that the band planned to release a parody of "Oh, Pretty Woman" and that Orbison and Dees would be credited as authors. The band would also pay the standard licensing rate established by the Copyright Office. Teifer denied her request, writing, "We cannot permit the use of a parody of 'Oh, Pretty Woman.' "[3]

But it's likely that Fine was not really asking Teifer's permission. She wrote him on July 5, 1989, and the group released *As Clean as They Wanna Be* ten days later with "Pretty Woman" as track 7, in between "Me So Horny" and "My Seven Bizzos." The album listed the song as " 'Pretty Woman' (parody of 'Oh Pretty Woman,' written by Roy Orbison and Bill Dees)."[4] Teifer's letter was dated July 17, so it certainly appears that the studio's rejection of the license took place *after* the album went on sale. The album was not a commercial success, selling only 250,000 copies (unlike its explicit counterpart, which sold 1.7 million). But "Pretty Woman" was on it, and even a single sale was enough for Acuff Rose to challenge its release.

That fall, Acuff Rose Music sued the group and their record label, Luke Skyywalker Records, in the District Court for the Middle District of Tennessee for copyright infringement. What stands out from this initial suit is that even if the legal complaint was copyright infringement, the underlying motivation seemed to be that Acuff Rose did not like what 2 Live Crew had done to the song. Acuff Rose argued that the lyrics "are not consistent with good taste or would disparage the future value of the copyright." It also insisted that the similarity of melody and lyrics between the two songs was substantial enough to reject the notion that 2 Live Crew's song was a parody. The group 2 Live Crew countered by moving for summary judgment, arguing that as a parody, their song was protected by fair use.[5]

TRANSFORMATIVE USES

Let's reflect on the problem that all courts handling this case had to discuss: Can a parody made for profit constitute fair use? Recall that the Copyright Act names four factors for courts to consider when assessing whether an action constitutes fair use:

(1) the purpose and character of the use, including whether such use
    is of a commercial nature or is for nonprofit educational purposes;
(2) the nature of the copyrighted work;
(3) the amount and substantiality of the portion used in relation to
    the copyrighted work as a whole; and
(4) the effect of the use upon the potential market for or value of
    the copyrighted work.[6]

The Supreme Court had never handed down an opinion on whether a parody constitutes fair use, but two opinions from the mid-1980s placed much emphasis on the fourth factor. Of course, an interpretation that placed a premium on commercial considerations is not surprising given the composition of the Court at the time, which was tending to be more conservative under the leadership of Chief Justice William Rehnquist.[7]

What is surprising, however, is that when 2 Live Crew's legal battle arrived at the Supreme Court, the Justices revisited this Reagan-era balance and, without realizing it, generated a ruling that would later become part of the Internet's legal backbone. The lower courts had disagreed on whether the group's use had been fair; the district court found that it was, but the appeals court had embraced mid-1980s Supreme Court reasoning and reversed that ruling. It had emphasized that there was excessive borrowing and that "the admittedly commercial nature of the derivative work . . . requires the conclusion that the first factor weighs against a finding of fair use."[8] 2 Live Crew's appeal to the Supreme Court therefore involved two questions: whether the commercial character of the parody excluded them from fair use protections, and whether the appeals court had given enough weight to the changes Luke Campbell had made to the song.

At the Court, firms and nonprofit agencies filed briefs affirming the interests we would expect. The National Music Publishers' Association sided with Acuff Rose Music, while the American Civil Liberties Union and organizations interested in free speech (such as the *Harvard Lampoon* and the PEN American Center) sided with 2 Live Crew. The stakes of the case were high in the sense that this was the first time the Supreme Court would rule on the relationship between parody and fair use, and a ruling against 2 Live Crew would have damaged artists' ability to mock others for profit. The oral arguments, however, made it very clear that the Justices did not see much controversy: a commercial parody could, in fact, be protected by fair use, and 2 Live Crew won by a 9–0 vote.

But when the Justices announced their decision, on March 7, 1994, they endorsed a theory of fair use that placed extraordinary weight on factor 1, the purpose and character of the use. Having digested decades of opinions on that factor, they explained that their main task was to decide whether the new work simply supersedes the original, of if it "adds something new, with a further purpose or different character, altering the first with new expression, meaning or message." They defined this use as *transformative*. But they added a crucial line, without citing any legal precedent: "the more transformative the new work, the less will be the significance of the other factors, like commercialism, that may weigh against a finding of fair use."[9]

This is why *Campbell v. Acuff Rose Music* is so important: the Supreme Court didn't just say that commercial parodies can be protected by fair use; it transformed a question about parodies into a general theory of what it means for a use to be "transformative," adding the stunning caveat that if a use is transformative enough, it will qualify as fair use. In 2 Live Crew's case, the limits of this reasoning were quite clear. Campbell had transformed "Oh, Pretty Woman" so much, the Court said, that his commercial intent was irrelevant. But the Justices didn't restrict their reasoning to parodies or even to written or musical works. Instead, they ruled that adding something new or altering the original in any medium could be enough to earn protection.[10]

This victory for free speech had an unexpected consequence in copyright law. A central tenet in current US copyright law—especially in Google's hands—is that a copyrighted work becomes something else when it is digitized and stored in a searchable database. The reasoning is that the work ceases to be a creative expression and becomes a technical object: it is modified in a way that gives it new meaning because the work is transformed from something meant to evoke a human emotion into an object meant to be processed by an algorithm. This is the central legal precondition permitting such things as image-based search engines, digital repositories of books, and even artificial intelligence systems that generate new text or art.

The company that established the link between the *Campbell* reasoning and search engines was not, surprisingly, Google but the Arriba Software Corporation, after a photographer demanded that the company take his pictures down from a search engine.

PHOTOGRAPHS AND SEARCH ENGINES

Leslie A. Kelly, from Huntington Beach, California, had been a photographer for decades before he first encountered the World Wide Web. More than two hundred magazines and books had published his work; his series documenting Amish people, the California Gold Rush, and real-life locations in Laura Ingalls Wilder's *Little House on the Prairie* books were especially popular. In the 1990s, Kelly realized that he could promote his work through websites and launched seven of them, each one dedicated to a different subject. He didn't sell his photographs through these websites, but he did include some of his photos to promote the many print outlets that were publishing them.

Kelly understood that his work as a photographer was a business. Operating under the corporate name of "Les Kelly Publications," he managed several simultaneous cross-licensing and cross-promotional endeavors. For instance, he published his Gold Rush photographs in a book, *California's Gold Rush Country*, which he promoted through a website, www. goldrush1849.com, and a companion book, *Traveling California's Gold Rush Country*. He also used some of these images on a different website, www.

showmethegold.com, which sold corporate retreat travel packages.[11] His licensing business included educational websites related to *Little House on the Prairie* (the television series based on Wilder's books) and formal associations with the Laura Ingalls Wilder Memorial Society and the South Dakota Department of Tourism, which were licensed to display his photographs on specific sites. He rarely licensed individual images, and he did not license his work to stock photograph services.[12]

In the spring of 1999, Kelly found many of his images available for free on a search engine called the ArribaVista, run by an Illinois-based company called Arriba Soft. Founded in 1997 by Michael J. Lyons, Arriba Soft offered "to simplify the lives of media professionals and individuals who use digital media" by "creating software tools that break down media barriers and allow access to media objects in a single environment."[13] One of its main products was a management program named Arriba Express, released in July 1998.

Arriba Express, which retailed for $149.95 after a fifteen-day free trial, allowed users to create, use, manage, and store multimedia files of all kinds. It was compatible with four hundred file types and eliminated the need to open separate applications for each kind of media users might have in their hard drives. According to Noëlle Leca, an executive from an e-commerce firm cited in the Arriba Express's press release, this program could reduce weeks' worth of image management labor to just a few hours. Its key time-saving feature was "Web Vac," through which a user could import files directly from a website or ArribaVista's thumbnails directly into Arriba Express for later use.[14] This feature even enabled auto-load of images from websites users preselected.[15]

The Arriba Express software became even more flexible when Arriba Soft released its second major consumer product: a free image-based search engine called Arriba Vista. This was a bot that would crawl the Internet searching for images. At each website, it would create a temporary copy of an image, create a thumbnail, delete the copy of the original, and store the thumbnail in its cache alongside keywords drawn from the surrounding text.[16] By clicking on the thumbnail—which was stored on Arriba Soft's servers—users could instruct their browser to display the original image directly from its host site, framed within the Arriba Vista display. The system blocked pornography and generated very flattering press. *PC Magazine* dubbed it the "power user's solution," and *Graphic Arts Monthly* celebrated how "the combination of Express and Vista provides an integrated media management solution." The *Los Angeles Times* praised its potential for family-friendly web browsing, and ZDTV included it in its list of "incredibly useful sites" and called it "the first Web search engine that searches exclusively for images."[17]

Arriba Express's copyright documentation is unavailable for research, but the website's terms of service included a playful note: "Arriba Soft Corporation encourages users of the Arriba Vista Image Searcher to use the site and have fun doing so. However, before using an image retrieved with the help of the Arriba Vista Image Searcher, you should contact the webmaster at the site of origin of that image and ask permission to use it. Remember, respect for individual intellectual property and the laws that protect that property is essential for the Web to remain a truly open resource for all users. When in doubt about materials you want to download, Ask First!!"[18]

In a separate page, titled "Disclaimer," the company offered a more incisive warning: "UNDER NO CIRCUMSTANCES SHALL ARRIBA SOFT CORPORATION OR ANY OTHER PARTY INVOLVED IN OR CONNECTED TO THIS WEB SITE'S CONTENT BE LIABLE TO YOU OR ANY OTHER PERSON FOR ANY DIRECT, INDIRECT, SPECIAL, INCIDENTAL OR CONSEQUENTIAL DAMAGES ARRISING [*sic*] FROM YOUR ACCESS OR USE OF THIS WEB SITE."[19]

The disclaimer page explained that Arriba Soft does not "copy or store the original," that it "merely provides a link to the location on the originating website," and that it does not grant or imply any licenses "to use these images other than for viewing." Copyright owners who did not want to have their images "indexed for retrieval by users" could notify the company by email and request that they be removed.[20]

KELLY GOES TO COURT

In the spring of 1999, Leslie Kelly found about three dozen of his photographs on Arriba Vista and asked the company to take them down. Finding that some images were still there a few days later, he filed a complaint at the District Court for the Central District of California. (The company had changed its name to Ditto, but I will keep calling it Arriba to align this discussion with the opinions' titles and their referencing in legal scholarship.)

Kelly retained Steven Krongold, a business and intellectual property lawyer in Costa Mesa, California.[21] Krongold knew that Arriba would counter with a fair use defense and that the court would be bound to consider the *Campbell* reasoning regardless of the parties' arguments. He made a very clever decision: claiming that Arriba was committing copyright infringement of both the full-sized images that it allowed users to download from Kelly's website and the thumbnails that it generated and displayed in response to users' search queries. There was a technical reason for this, of course—Arriba created the thumbnails, not Kelly—but it also had the strategic advantage of forcing the court to issue two different rulings, making a partial victory possible.

Arriba Soft's use of the name "Web Vac" gave Kelly a key comparison to plead his case: Arriba Vista's caching was like a vacuum cleaner sucking up items from the web. Arriba Express, according to the complaint, "incorporates software that will collect or 'vacuum' all images from a targeted web site."[22] The complaint explained: "The images vacuumed from other websites and copied to defendant's www.arribavista.com are displayed as 'thumbnail' copies which users can download to their own computers using the free image search engine by simply right-clicking the mouse button."[23] If users also owned Arriba Express, they could even download the full-sized image directly from the website simply by clicking a button embedded into ArribaVista.

The key assumption behind this description of Arriba Vista—that Kelly's copyrights over an image extended to any thumbnails of them—was tied to a commercial consideration: that Arriba was improperly profiting from Kelly's work and costing him advertising profits. The site displayed banner ads in its results page and in the new page that would open once a searcher clicked on one of the thumbnails. This new page displayed the original image through in-line linking, offering a window into the source website framed within Arriba Vista itself. This allowed the searchers to see an image without also seeing the advertisements that Kelly had put on his site. All this is to say that the complaint charged Arriba Soft Corporation with copyright infringement, violation of the DMCA (for failing to take down images), and unfair competition.

Arriba hired Perkins Coie, one of the most powerful global law firms, to represent it. The lead counsel was Judith B. Jennison, who had developed extraordinary expertise in software law from participating in such major cases as *Atari v. Nintendo* and *U.S. v. Microsoft,* which was still unfolding.[24] She and her colleagues developed a bold two-part strategy. One part was a technosocial argument: search engines are an emerging technology with immense potential benefit to the public. Second, they argued that Arriba Soft's use of thumbnails qualified as a transformative under *Campbell* because the thumbnails linked to the source webpages and were also linked to keywords in the engine's index; and that Arriba's display of full-sized images was not an infringement because the software merely allowed users to download the images from the original server that hosted them. Should the court agree on both fronts, it would be easier to argue that Arriba was protected by fair use, even though (as the lawyers would later admit), Arriba "copied Kelly's images in their entirety."[25]

Judge Gary Taylor, appointed by George H. W. Bush, granted summary judgment in Arriba's favor, following the reasoning in *Campbell v. Acuff Rose Music* almost to the letter. He wrote that while Kelly's photographs are "artistic works used for illustrative purposes," Arriba's "visual search engine is designed to catalog and improve access to images on the Internet." In other words, the "character of the thumbnail index is not esthetic but functional; its purpose is not to be artistic but to be comprehensive." Arriba had "swept up" Kelly's images "along with two million others available on the Internet," but all in the spirit of providing users "with a better way to find images on the internet."[26]

Taylor's ruling on the full-sized images, again in Arriba's favor, looks to me like plain pro-corporation bias. Since this was a new kind of business and the technologies involved were not fully developed, he thought it best to side with Arriba. In his words, Arriba's "purposes were and are inherently transformative, even if its realization of the purposes was at times imperfect. Where, as here, a new use and new technology are evolving, the broad transformative purpose of the use weighs more heavily than the inevitable flaws in its early stages of development."[27]

Here Taylor was referring to Arriba Vista's image attributes page, which users could access by clicking on the thumbnail. This page allowed users to view and download full-sized images without having to visit the original webpage, so its connection to the larger purpose of "finding and organizing Internet content" was nebulous. The judge recognized that this page "somewhat detracts from the transformative effect of the search engine." But since Arriba Soft had recently removed that page from its search engine and created a new search interface as part of its rebranding and name change from Arriba Vista to Ditto.com, Taylor concluded that he no longer needed to worry about the attributes page. "It is more appropriate to consider the transformative purpose," he wrote, "rather than the early imperfect means of achieving that purpose."[28]

ASK FIRST

Kelly's stunning defeat at the district court cemented his case's status as a proxy for the concerns of artists worldwide. In a statement published at NetCopyrightLaw.com, he lamented that the opinion "clearly sends the wrong message to artists who hope to use the Internet as an important new market for their services." By marketing their services as free, companies such as Arriba Soft were adding "to the already widespread and unfortunate belief that anything and everything on the Internet is 'free' for the taking."[29] In his view, the court was favoring a multimillion-dollar company's desire to exploit the web for profit over artists' efforts to use their intellectual property rights to make a living. The Graphic Artists Guild and the National Writers Union—both part of the United Auto Workers—had helped him with his litigation costs and were ready to aid his appeal, as was the American Society of Media Photographers.

At the Graphic Artists Guild, Kelly's battle meant the escalation of a long-standing problem—artists' work being treated as a free good. The guild had long had a code of fair practice for the graphic communication industry that was designed to "promote equity for those engaged in creating, selling, buying, and using graphic arts."[30] Created in 1948 and often updated, the code deemed electronic rights "separate from traditional media" and emphasized that rights to artwork are "always vested in the hands of the artist" unless otherwise specified in writing. The guild had even partnered with the Copyright Clearance Center to launch a campaign, Ask First, aimed at raising awareness of intellectual property rights. Part of the campaign's website read, "We appreciate your desire to use our images. Even more, we are flattered and complimented. But for a number of reasons, artists may not want to have their images used in any way, including agency representations. And any use, including 'comping,' implies value that is worth some compensation. Contact the creator before copying."[31]

Ask First highlighted two situations in which artists were particularly vulnerable: when they submitted portfolios for job applications and when their images were included in talent sourcebooks and directories. The Kelly case seemed like an extreme version of such situations. Not only was Arriba making an artist's work widely available, but every person who downloaded the photographs might use them without his consent. (Visit the link in the notes to see Ask First's logo and website, which I could not reprint here.)

Kelly's case generated spirited debate among bloggers. Dana Blankenhorn, a professional Internet reporter and frequent contributor to the online news outlet ClickZ, called the suit "The Dumbest Lawsuit in Web History." He presented the case as the "latest twist" in the long effort at "declaring the web illegal." According to Blankenhorn, Kelly aimed to "make search engines illegal" because one of them was stealing his pictures. The suit was especially "stupid," he thought, because Kelly could have simply placed his images behind a firewall or password protection, or he could have just mentioned his pictures instead of posting them. "If you don't like the way the web works," he concluded, "don't put your stuff on a website."[32]

Blankenhorn's commentary generated very strong buzz. Readers sent emails to ClickZ calling it unforgivable, shameful, a travesty, and a number of less savory things. Julia Ptasznik, an instructor at the Fashion Institute of Technology, responded with a piece titled "This Pisses Me Off" in the trade publication *Visual Arts Trends*.[33] One of her students, she wrote, had begun a class presentation by asking her classmates, "How would you like it if an image posted on your site was taken without your permission and used by another website for commercial purposes, without any notice or payment to you? Well, that's exactly what happened to photographer Les Kelly."[34] Ptasznik warned that this case was about more than just photographers' work: it would set a precedent for all copyrighted material by delineating the rights that automated search engines had over creative people's work. She was especially worried about a new assumption that seemed to be developing: that creators who placed their work online were implicitly granting web surfers the right to do whatever they wanted with it.

Ptasznik's readers shared this concern. Michael Anderson, an artist in California, warned that creators who used the Internet were putting their goods within arm's reach of anyone in the world. His response was to avoid posting online any works he actually considered valuable, but if all creative people did this, the Web would devolve into "a no-man's-land of ASCII text and consumer-marketing graphics (#%$^!)." Another artist wrote that unless "the good side wins," creators "will just have to hold on to our hats and weather the storm of thieves and truants of the www." Blankenhorn himself responded with a brief note implying that if text could be quoted and music could be sampled, then visual artists should not mind thumbnails.[35]

Creators across several art forms saw the case in terms of whether artists would have enough control over their works in the online world. Like virtually every lawyer in copyright circles, they would closely follow Kelly's appeal to the Ninth Circuit Court of Appeals.

THE TWO KELLYS

Kelly filed his appeal on July 19, 2000, with financial backing from the American Society for Media Photographers and the Graphic Artists Guild.[36] Arriba would continue to be represented by Perkins Coie, with Jennison in the lead, but Kelly now had heavy-hitting attorneys. Leading the charge was Charles D. Ossola, the head of the Intellectual Property and Technology Practice Group at the DC firm of Arnold & Porter. Ossola was also the outside counsel for the American Society of Media Photographers.[37] Backing him up were two well-known corporate lawyers and Kelly's original attorney, Steven Krongold.

Ossola's strategy was to argue that the district court had erred in counting Arriba's handling of the images as transformative. This argument depended on two claims. First, adding content such as hyperlinks to a copyrighted work is not enough to claim a transformative use. The main example of this was a case decided in 1997 in which the Ninth Circuit had ruled that a news agency's unauthorized broadcasting of a copyrighted video (depicting the brutal beating of Reginald Denny) was not transformative just because the agency added a voiceover.[38]

The second claim was that the district court had been improperly swayed by discussions of the public benefit of Arriba's actions, favoring these benefits over a fair use analysis. Ossola noted that the search engine companies Altavista, Yahoo!, and Google had filed a joint brief urging the court to follow the "conventions that govern the web, whatever those may be," but arguing that a ruling in Kelly's favor would make it impossible for the web to function. Requiring all websites to get licenses to link or display images would paralyze the web and jeopardize its future.[39]

In 2002, Judge Thomas Nelson handed Kelly a partial victory: he ruled that "the creation and use of the thumbnails in the search engine is a fair use" but that "the display of the larger image is a violation of Kelly's exclusive right to publicly display his works."[40] The ruling on thumbnails was firmly grounded in *Campbell*. Nelson justified it by noting that thumbnails are "much smaller, lower-resolution images" that serve a different function from high-resolution images. "Kelly's images are artistic works used for illustrative purposes," Nelson wrote. "[They] are used to portray scenes from the American West in an esthetic manner. Arriba's use of Kelly's images in the thumbnails is unrelated to any esthetic purpose. Its search engine functions as a tool to help index and improve access to images on the internet and their related web sites."[41]

Nelson's assumptions about the nature and uses of thumbnails were not well justified. He explained, for example, that "users are unlikely to enlarge the thumbnails" and use them for illustrative or artistic purposes because the thumbnails have lower resolution.[42] There was no evidence for this, just Nelson's belief that a low-resolution image ceases to have artistic uses.

But he did recognize that this argument did not apply to Arriba's full-sized images, and the company had not addressed them explicitly in its fair use argument. Here, the judge saw a clear-cut example of copyright infringement, with all fair use factors weighing in Kelly's favor. Yet his finding was grounded on a serious misunderstanding of how Arriba worked. Because screenshots showed Kelly's image embedded in a page by Arriba, Nelson assumed that Arriba was the one displaying (and perhaps even hosting) the full-sized photographs.

This assumption was wrong. Arriba had *framed* the image within its page—that is, it showed the image that Kelly's website hosted without storing it itself. The image Arriba users saw, when they wanted to see Kelly's work, was Kelly's own website placed within an Arriba frame. The way this occurs in a website is by placing the URL *for the image* in Arriba's frame: the image shown in the frame is actually Kelly's site.

If I may use a comparison of my own: this is like opening a window so that you can see the platform. You have probably become accustomed to this kind of arrangement. For example, when a website shows a tweet or Instagram image, it does not necessarily post a screenshot from the platforms. Instead, the website's content creator selects an area on their site where content from the platform will be visible. Try it yourself: the next time you see an embedded tweet or Instagram post in a new story or Buzzfeed listicle, click on it. (Be careful, please, and do so at your own risk.) If you are taken to the original post, it is possible that you went out the window and into the platform.

Nelson's opinion caused an uproar among Internet companies and at the Electronic Frontier Foundation. A few days after the opinion came down, the EFF filed a brief asking the Ninth Circuit to reconsider. Its press release charged that the ruling "threatens to make all linking on the World Wide Web a copyright infringement" and that Nelson had created "a novel legal rationale" with his holding that "linking to a website without permission infringes the public display rights of the website owner." Fred von Lohmann, a senior intellectual property attorney at the EFF, wrote on the foundation's website that "extending copyright law to all linking on the Web would be a nightmare" and that "a whole new flurry of lawyer letters would chill linking on the Web, striking at the heart of free expression on the Internet."[43] Google followed with a brief requesting clarification, claiming that the ruling would cripple search engines generally.

The EFF brief explained that the court had misunderstood how the technology works. Its explanation of how linking works has become standard in these kinds of cases, so it is worth quoting at length:

> A user employs a "browser," such as Internet Explorer, in order to send requests to these Web servers. Links provide a shortcut for sending such requests. For example, if a Jane Doe clicks a link to the EFF homepage, her browser will send a request to EFF's server asking, "send me the document you have stored at the address www.eff.org." This is identical to what would have occurred had she typed or pasted the URL "http://www.eff.org" into her browser. The EFF server would then respond by sending a set of instructions, which Jane's browser will use to assemble and ultimately display the EFF homepage.

The browser's initial response to this request specifies the layout of the page and supplies additional URLs corresponding to images, texts, and banners that could come from anywhere on the web. Without her knowledge, Jane's browser "responds to each of these URLs just as if she had clicked on a link for each of them—it sends requests to each of the servers specified." Her browser then "assembles all the materials received and displays the EFF homepage."[44]

Kelly's original allies and some additional interest groups—including the American Society of Media Photographers, the Authors Guild, and the National Music Publishers' Association—argued that the court's ruling did not threaten linking. But their understanding of linking was limited to showing URLs of other websites. They maintained that "merely linking to another's site *is* permissible. But going beyond that to display full-sized images from that site . . . goes too far."[45]

By the time Arriba's new parent company (Sorceron) appealed the ruling on March 23, 2002, the question was not just whether Arriba had infringed on Kelly's copyrights by displaying full-sized images. It was whether copyright law allowed a website to frame another site's content. But the Ninth Circuit did not answer this question explicitly. In 2003, it upheld fair use protections for thumbnails and sent the case back to the district court for reconsideration of full-sized images. Sorceron went out of business soon thereafter, and on March 17, the district court entered a default judgment in Kelly's favor.[46] He had won, but the battles over how search engines handled copyrighted material were just beginning.

Almost every year, when I teach my "Digital Lawsuits" class, the story of how 2 Live Crew ended up involved in a landmark Supreme Court opinion always makes my students smile. It is a joy to teach. I play samples from Roy Orbison's "Oh, Pretty Woman" and then ask the class to brace themselves for the Crew's imaginative remix. Students' reactions range from laughter to shock before becoming puzzled looks as I explain *Campbell*'s entirely unexpected impact on the online world. That is the beauty of teaching law as a historian: I get to tell them the stories you're reading right now, to see their faces as they realize how individual threads of legal thought are woven into their online world.

But next time I teach 2 Live Crew's saga, the bigger picture will be quite different because it will have to involve Andy Warhol, Prince, and a photographer named Lynn Goldsmith. In 2023, the Supreme Court decided the case of *Warhol v. Goldsmith*. Back in 1981, Goldsmith had taken a photograph of the musician Prince. Three years later, she licensed the image to *Vanity Fair* magazine for $400, so that Warhol could make a silkscreen illustration of Prince that would be published only once. Thirty-five years passed, and in 2016, after Prince's death, *Vanity Fair*'s publisher (Condé Nast) licensed Warhol's silkscreen and used it as the cover for a special issue of a magazine. I can't include these images in this book to avoid being charged with infringement—which would be very likely, given the situation—but you can see plenty of images by typing the term "Lynn Warhol Prince" into Google Images.

In May 2023, at the Supreme Court, the question was whether the Warhol Foundation was protected by fair use when it licensed the image to Condé Nast. The foundation argued that the artist's use of the photograph was transformative *because* the silkscreens convey a meaning or message different from the photograph's. This would mean, in turn, that it had the right to license the silkscreen without Goldsmith's permission. In a 7–2 opinion written by Sonia Sotomayor, the Court disagreed. Justice Sotomayor explained that the first fair use factor is not solely concerned with *whether* there is new meaning or message—that is, on whether a use is or isn't transformative. Instead, the question is "a matter of degree, and the degree of difference must be weighed against other considerations, like commercialism."[47]

This is the single most significant opinion on transformative uses since *Campbell*. It's a simple but powerful statement, that a use is not fair just because it is transformative. There is more to tell about how the notion of transformative uses in the style of *Campbell* became central to online copyright. Had this book been published in 2022, that would have been the end of it. But the *Warhol* opinion reminds us that the dense web of legal reasoning that sustains online copyrights is also very unstable. As you read on, ask yourself: How might these stories have unfolded differently had *Warhol* been issued twenty years ago?

CHAPTER EIGHT

Bad Poems and Free Porn

IT'S OCTOBER 7, 2023, I'm sitting on a reclining chair next to my chaotic home library revising the chapter you are reading right now, and I need to consult Siva Vaidhyanathan's classic book *The Googlization of Everything*. I'm sure I own a physical copy of it because I bought one during graduate school, but I must have lost it when my husband and I moved to Sacramento.

I type the book's title into the Google search bar and find some promising leads: the book's official listing at the University of California Press, a few reviews, and a link to the full book on the academic database JSTOR. I also find a few dozen pages of the book on Google Books, but I'd rather have the whole thing. There is, of course, an easy solution: visit an illegal PDF repository called Z-Library, which keeps getting shut down and reemerging on other servers. My first book was there two days after it was published, so I have a permanent grudge against the site even though many of my graduate students use it every week.

I log into my university's network, and I don't have access to the JSTOR copy. That happens sometimes: university presses sell access rights to several databases, and universities pick and choose which ones they contract with. So I type the book's name into our library's website, and in less than a minute I have a PDF of the entire thing. It has taken me several steps to avoid infringing on Vaidyanathan's and UC Press's copyrights.

But now I have the book, so I can keep writing. Vaidyanathan used the term "Googlization" to describe how "Google has permeated our culture." He documented how, in just a dozen years, Google had given everyone access to an unprecedented number of resources. The name "Google" had become both a brand name and a verb: to "Google" something means to search for it using the company's search engine. Vaidyanathan's book called for us to stop thinking of Google as a benevolent corporation, realize that we are the products it's selling, and understand that we're all caught up in Google's "infrastructural imperialism."[1]

I knew all this when I started looking for Vaidhyanathan's book, yet I still went to Google as the first stop in my search for it. There are other search engines, but I dislike some of them (Yahoo!) and disdain others (Bing). The reason is that, as Vaidyanathan predicted, Google is the default infrastructure for all my professional needs; my calendar, email, surveys, research, photo archives, and academic collaborations are all stored under various Google usernames.

The reason I needed to find Vaidyanathan's book is that I wanted to explain how, for the past fifteen years, I have been witnessing the Googlization of copyright law and how Google's efforts to protect itself from liability for copyright infringement have transformed the law itself into a tool that favors universal access over individual rights. I wanted to show how copyright law has become a central force in Google's infrastructural imperialism, and I needed a page number. The *kind* of legal reasoning that has allowed this to happen—the notion that a creative work becomes something else when it is handled as data—was not invented by Google. It developed slowly over the years, as courts handed down opinions on what it means to transform a creative work. But Google and other digital giants have been champions of this reasoning, increasingly prompting courts to favor their corporate needs over those of online content creators.

This chapter tells the story of two court cases: one involving a lawyer who worried about Google's ability to store his poems on its servers, the other about a very litigious pornographer who wanted to prevent potential clients from finding his company's product on Google Images. Had Google lost either of these cases, text and image searches might now be legally untenable. But its victories paved the way for companies to store and process massive amounts of creative works, regardless of the creators' desires. The result, in a great irony for a copyright regime, is that creators have to opt *out* of having their work shown and stored without authorization.

GOOD TEA AND GOOGLE BOTS

In the early 2000s, a lawyer named Blake A. Field decided to test his luck in a lawsuit against Google. His online research had taught him the basics of how online search engines worked: the engine's bot would crawl through a page, create a cached version, and use that version to respond to users' queries.[2] He had also learned that Google automatically created links to cached pages and included them in search results unless the page's publisher explicitly asked the engine not to do so, either by including a "no-archive" metatag on its website or following a takedown request available through Google's site. Early in 2004, Field created a website for inclusion into Google's search engine.

Over the next three days he wrote fifty-one short works and posted them for free at www.blakeswritings.com. As with song lyrics, I must cite very little from his poems to avoid a copyright claim, but you can read the full poems at the links in the notes. Many of the poems were stylized opinions about everyday subjects. For instance, "Good Tea," a poem with six stanzas, opened with, "It's a truly rare find, a good tea."[3] Other poems offered humorous commentary on many topics. For instance, "Antiperspirant" opened with, "It's a fact that I sweat like it's my job."[4]

Copyrights offer protection regardless of a work's merits. Field registered copyrights for all his work soon after publishing the site.[5] Available documents suggest that he tried to purchase advertisements for his site, but he did not charge any fees to view his works. He also took special care to ensure that search engines indexed the site: he added a robots.txt file allowing all crawlers to index his site, chose not to include a no-archive tag, and submitted his site to Google to expedite the Googlebot's arrival. He would later tell attorneys that his goal was to get the site indexed as widely as possible by allowing "the maximum number of crawlers."[6]

The Googlebot and other web crawlers eventually visited his site, indexed it, and linked to the cached page in search results. If Field had wished to do so, he could have done many things to ask Google to remove his page from the cache. For instance, Google offered a form that publishers could use to ask that their content be removed. He could have added a robots.txt file instructing the Googlebot not to archive the site the next time it crawled through. Rather than doing any of these things, however, he filed a complaint against Google at the District Court for the District of Nevada on April 6, 2004, and neglected to inform Google.[7]

The complaint focused on "Good Tea." It noted that the cached page bore his website's copyright notice, "© 2004 Blake A. Field," and claimed that for everyone who visited the cached website, Google created a copy of "Good Tea" and distributed it without Field's authorization. Attached to the complaint was a printout of the cached page itself, which someone—perhaps Field's lawyer—had obtained by clicking on the cached link from a computer in Las Vegas, Nevada, where Field resided.[8] Field claimed that Google's reproduction violated his rights to reproduction and distribution and that the company had done so in a way that was "willful, intentional, purposeful, and in reckless disregard of and indifference to" his rights."[9] This had caused him to "sustain substantial injury, loss, and damage to his rights," and he asked the court to enjoin Google from distributing his story and compensate him with no less than $50,000 for "Good Tea" and each of the other fifty stories—a total of $2,550,000 in statutory damages.[10]

Ten days later, an in-house attorney at Google, Michael Kwun, sent Field a letter stating that Google had obtained his complaint. Kwun explained that this was the first time Google heard about Field's desire not to have his pages cached and that the company had "acted promptly to remove all pages" from its cache. Kwun's letter directed Field to www.google.com/remove, which "provides straightforward guidelines" for informing Google that "unlike the majority of web site publishers," he did not want Google to cache his pages. "I assume that stopping our provision of the Google cache of your site to our users was the purpose of your lawsuit," Kwun wrote. "Now that we have removed your site from the Google cache, I assume you will dismiss your lawsuit."[11]

When Field did not withdraw his complaint, Google responded by seeking declaratory judgment based on three claims: that Google's caching constituted fair use, that Field had given Google an implicit license to cache the works, and that Google was protected by section 512 of the Digital Millennium Copyright Act. At the heart of the response was the implication that Field was misrepresenting his own technical choices and creative ambitions in order to trap Google. The company alleged that Field had created his works only to launch a lawsuit and intentionally ignored the company's clearly outlined procedures to exclude sites from the Googlebot's crawl.[12]

BLACK BOX BOTS

As the case progressed, Google delivered many declarations and documents meant to portray its search engine as a black box that requires very little human intervention. One of its product managers, Bill Brougher, emphasized that the engine works by "compiling an index of the content available on accessible Web sites and querying this index rather than querying the billions of different Web pages."[13] Brougher explained: "Google—like other search engines—uses an automated software program (also known as a web crawler, spider, or Googlebot) to obtain copies of publicly available Web pages. It would be impossible for Google to locate and index all of the Web pages manually. The Googlebot, like other users of the Web, obtains copies of Web pages by sending requests to the server for the originating Web site and receiving the requested content in response."[14]

In this first stage, according to Brougher, the Googlebot was like a human user in that it obtained websites by asking servers for content. But unlike a human user, the bot was gathering content not to understand or appreciate it but for Google to index it. Brougher continued: "Google analyzes a copy of each Web page it receives from the originating Web servers and stores the copies in Google's cache through an automatic technical process. When a user submits a query, the Google Search Engine searches the Google system cache and quickly returns a search results page containing links to a list of Web pages within Google's index that are relevant to the user's query."[15]

Brougher's words must have been carefully and strategically chosen because they imply something with great legal significance: that cached copies are temporary by-products of automated processes. According to Brougher, "Google maintains a copy of a given Web page in the Google system cache only until the Googlebot next visits the particular Web page. This allows the Googlebot sufficient time to make it through the billions of Web pages and return to the particular Web page for re-indexing."[16] Perhaps the bot was like a human in that it requests information, but the comparison ends there. At Google's servers, the websites did not have the same meaning, uses, or functions they would have had if a user turned them into PDFs to browse them later. The bot was not amassing information as a Napster user did with songs. It was assembling a database that allowed for the search engine software to function.

Since 1998, users had been able to access many of these cached sites by clicking on a small link found underneath the link for the original site.

Google provided these links "through an automated technical process" to allow users to "see the copy of the Web page used by Google Search."[17] Users who clicked on these links saw a copy of the cached page automatically modified in response to their query. At the top of the page was this disclaimer: "Google's cache is the snapshot that we took of the page as we crawled the web. The page may have changed since then."[18] The page also showed the day of the Googlebot's last visit and highlighted any search terms the user had entered. It did not display any advertisements.

The Googlebot did not cache pages with traditional access controls such as requiring passwords or registration prior to accessing content. But the company also offered a coding guide to remove publicly available content from the Googlebot's crawl or snippets from the search engine's results page. This guide followed a set of informal standards developed by a global community of developers, who self-identified as "robot authors and other people with an interest in robots."[19] For instance, the owner of a site called yourserver.com could add a file with the following text to their root directory:

```
User-agent: Googlebot
Disallow: /
```

This would tell the Googlebot not to crawl through the page to cache it. Google's guidelines for content removal said that publishers had to opt out of the crawl because the "comprehensiveness" of its search results was "an extremely important priority." The company claimed that its commitment to "providing thorough and unbiased search results" meant that it could not "participate in the practice of censorship" by blocking sites from its end. Of course, sites that were spamming Google's index or that were required by law to be taken down were also excluded.[20]

Despite these opt-out mechanisms, Brougher maintained, Google's cached pages offered "a number of benefits." First, users could read cached pages for originals they might not be able to access. This could happen, for example, if a website had too many visitors at the same time, if the page had been deleted, or if their national governments or Internet service providers blocked access to it. Second, the cache had an "archival functionality." It allowed users—including researchers, teachers, and journalists—to compare live versions of a site with a previous version that might have undergone substantial revision. Third, the cached page made reading easier by highlighting terms that helped users "determine why a particular page was deemed responsive to their search query."[21]

This was part of Google's ambitious effort to argue that whereas siding with Field could have an extremely detrimental impact on the public, siding with Google would have no negative impact at all. The company relied heavily on the testimony of John R. Levine, a prominent computer scientist, antispam advocate, and author of *The Internet for Dummies*, to make this point. Levine echoed Brougher on technical matters, but he also emphasized that "the uses of the Google system cache do not negatively impact demand for the original web page."[22] Cached pages were complements, not substitutes, to the original. In any case, use of these pages would have no "appreciable economic impact on the original website" because the cached page presented the advertisements that the website included when the Googlebot crawled through it.[23]

Levine used the Internet Archive (Archive.org), a nonprofit organization that grants free access to digitized materials, to argue that Field had explicitly granted permission for search engines to crawl through his page. He visited an archived version of Field's homepage and found the following metadata:

> meta name="keywords" content="short stories story writings
> writing blake field I blame anne murray box of macaroni band
> t-shirt antiperspirant bass slide broken headphones dogbait drive
> eyedrops filthy comforter left foot ink cartridge lotions oil change
> black cow ronald reagan will never die oil change room cleaning"[24]

The metadata contained keywords that search engines could use, so Levine concluded that whoever had created the homepage knew what these keywords were for and had prepared the list with the clear intent of allowing search engine bots to crawl through the page. Even if this code hadn't been there, though, publishers who did not include the code necessary to keep bots off their pages were effectively granting permission for this to happen. Without such "implicit permission," Levine concluded, "search engines simply could not operate."[25]

SAVING THE SEARCH

Google's strategy—positioning Field as a troll suing Google over automatic processes that he had implicitly allowed—worked in every respect. The black boxing of the search technology enabled the court to reason that the end user, and not Google, was the one who caused Google's servers to deliver cached copies of Field's pages. According to Judge Robert Jones, when "a user requests a Web page contained in the Google cache by clicking on a 'Cached' link, it is the user, not Google, who creates and downloads a copy of the cached Web page. Google is passive in this process. Google's computers respond automatically to the user's request."[26] Without a user's request, Jones wrote, no copies would be created and sent. Google was acting in an automated, nonvolitional way that absolved it from Field's accusations of infringement.

Jones ultimately favored all of Google's arguments and granted it summary judgment on all fronts. His reasoning was grounded on court opinions that had been developing since the 1990s. In the Netcom Scientology case, for example, the District Court for the Northern District of California had explained that any theory of copyright infringement for the online world would need to avoid creating liability for the chain of servers that connected the user who posted something to the one who downloaded it. Otherwise, courts would needlessly create liability for every server "in the worldwide link of computers" that transmits an infringing message.[27] More recently, in 2004, in the case *CoStar v. LoopNet,* an appeals court had ruled that the DMCA's safe harbor provisions were "a floor, not a ceiling, of protection" for ISPs. It held that ISPs do not "copy" material when they are "passively storing material at the direction of the users to make that material available to other users upon their request."[28] The *CoStar* opinion affirmed that the automatic copying, storage, and even transmission of copyrighted materials did not make ISPs liable for infringement instigated by others. In these cases, even if ISPs were liable for vicarious infringement, the DMCA provided a safe harbor.

Jones thought that the whole ordeal was a setup. He concluded that "Field decided to manufacture a claim for copyright infringement against Google in the hopes of making money from Google's standard practice." The judge's turn-by-turn account of the events that led to the conflict emphasized all the ways in which Field took actions explicitly to target Google. For instance, Jones wrote that Field "created a robots.txt file for his site and set the permissions within this file to *allow* all robots to visit and all the pages."[29] But *Field v. Google* became a major turning point in copyright law because Jones wrote an opinion addressing both the legal implications of this behavior and the broader copyright issues relevant to the case.

Jones ruled that Google's caching of websites was fair use by building on the theory of transformative uses found in *Campbell v. Acuff-Rose Music* (the 2 Live Crew case) and in *Kelly v. Arriba* (on thumbnails). Unlike a poem, Google's cached links allowed "users to locate and access information that is otherwise unaccessible." The links let users "detect changes that have been made to a particular Web page over time" and "understand why a page was responsive to their original query." The cached links may have had the exact same text as the original, but they had different fonts and clearly displayed links to the original. These uses and features led Jones to conclude that Google's caching was highly transformative in the *Campbell* sense: that instead of trying to replace the original, they added "something new, with a further purpose or different character, altering the first with new expression, meaning, or message."[30]

And there we have it again: the argument that once a creative work becomes part of a technical system, it becomes something else—and that incorporating that work into the system is an action protected by fair use. Jones wrote: "Assuming that Field intended his copyrighted works to serve an artistic function to enrich and entertain others as he claims, Google's presentation of 'Cached' links to the copyrighted works at issue here does not serve the same functions."[31]

In 2024, as I finished writing this book, Google announced that it would no longer make or grant access to cached versions of a site. The company's announcement on X made caching sound very old-fashioned. Google's X account for search technologies posted: "Hey, catching up. Yes, it's been removed. I know, it's sad. I'm sad too. It's one of our oldest features. But it was meant for helping people access pages when way back, you often couldn't depend on a page loading. These days, things have greatly improved. So, it was decided to retire it."[32]

Caching may no longer be front and center at Google's interface, but the reasoning that allowed it continues to be foundational to the relations between information technology and copyright. Authors and publishers are grappling with it today, because generative AI technologies such as ChatGPT were trained on online texts without their writers' consent. These systems process millions of entries of whatever medium they are designed to output, so that they can respond to prompts in whichever way best matches what they have "learned" from this training.

ChatGPT was trained on millions of texts found across the web, from blog posts and Wikipedia articles to books and message boards. It may seem that ChatGPT is chatting with you, but in reality it is constructing strings of words one at a time. It does not actually understand what you ask, but it can compute (based on more than one trillion parameters) which sequence of words is most likely to be a response based on the texts it has processed. You can see this for yourself by asking ChatGPT to generate a list of articles on any topic. Some of the articles it lists will be real (most likely because they are very popular), but many of them will be fictional works that sound like the perfect match for what you asked. (Unfortunately, more than one student has discovered this too late, after submitting a paper.)

Psychologist Steven Pinker did a brilliant little exercise to illustrate ChatGPT's inability to understand in a more human sense. He asked ChatGPT, "If Mabel was alive at 9 a.m. and 5 p.m., was she alive at noon?" In an article for the *Harvard Gazette,* he reported that the system responded: "It was not specified whether Mabel was alive at noon. She's known to be alive at 9 and 5, but there's no information provided about her being alive at noon." Pinker concluded that "we're dealing with an alien intelligence that's capable of astonishing feats, but not in the manner of the human mind."[33]

The miracle of public-facing generative AI such as ChatGPT was made possible in part by the Googlization of online copyright. Artificial intelligence developers have been training their systems on millions of creative works on the well-founded legal assumption that introducing a work into a database or search engine transforms it into something else: a technical object meant to enable a technological system to function, as opposed to a creative expression meant to evoke human emotion. *Kelly v. Arriba* and *Field v. Google* partly justify this assumption, but truly understanding its logic requires us to turn to pornography once again.

GOOGLING PORN

In the late 1990s, Norman Zada's career took a quick turn from financial management to soft-core pornography. Zada (born Zadeh) was, by all accounts, a financial mogul. He was the son of Lotfi Zadeh, a computer scientist best known for coining the term "fuzzy logic" in the mid-1960s. After earning a doctorate in operations research from the University of California, Berkeley, in 1970, Zada joined the staff of IBM's Thomas J. Watson Research Center. For the next decade, he taught applied mathematics, finance, and operations research at major universities across the country. He became a prominent money manager and founder of investing competitions such as the US Trading Championship and the US Investing Championship.[34] He was also an accomplished gambler; his book *Winning Poker Systems*, from 1974, remained a popular text for decades.

In the mid-1990s, after cofounding a hedge fund and raising more than $150 million from four hundred investors, Zada decided to invest some of his fortune in the adult entertainment industry.[35] His entry into porn became the stuff of legend among financial managers. After abandoning the idea of opening a strip club, Zada was struck when a woman he was dating wanted to be featured on the *Playboy* site but was rejected. To help her feel better, the story goes, Zada decided to create *Perfect 10*, an adult magazine focused exclusively on women who had not had plastic surgery.[36] *Perfect 10*, it seems, was more about Zada's desire to soothe his partner's feelings than about challenging Eurocentric conceptions of beauty, even despite his claim that he aimed "put out something that raises the standards." He told the Associated Press that he hoped to recapture the "classy days" of *Playboy* before it became "implanted and raunchy." "I've been to strip joints and fallen in love," he said. "And then you realize they're not real, and it's heartbreaking."[37]

Perfect 10 was not aimed at turning a quick profit. To make this point, and to flaunt his wealth, Zada told the Associated Press, "Look if I lose $500,000 a year on *Perfect 10*, I'll be happy. . . . It's not about that."[38] Media outlets reported that he had spent over $2 million of his own money to launch *Perfect 10*—a new magazine that, according to *US News*, would feature "relatively tasteful topless photographs and meaningful interviews on subjects like the downfall of communism and the turn-off of dandruff."[39] The magazine sold for $6.95 a copy, initially to fewer than five hundred subscribers, and advertisers were hesitant to buy ad space. At the time, the adult magazine industry had more than two hundred publications featuring female nudes. *Playboy*, the dominant publication, had more than 2.5 million subscribers and regularly attracted celebrities such as Jenny McCarthy.

This loss-tolerant approach became a problem when *Perfect 10* grappled with the decline of magazine publishing in the early 2000s. Zada coped with this shift in the landscape by abandoning print issues in 2007 and transitioning to a subscription-based website that charged $25.50 a month. To prepare for this online move, the company invested about $36 million, hiring eight hundred models and creating 3,300 images for distribution through perfect10.com. It also experimented with an alternative business model, granting a UK firm called Fonestarz Media a license to distribute reduced-size images for use in cell phones. This was Perfect 10's only licensing arrangement, and it facilitated the sale of about 6,000 images per month.[40]

Perfect 10's internal records are unavailable, but documents produced during the company's many legal battles suggest that it was never self-sustaining and that by 2015 it had lost at least $50 million.[41] Rebekah Chaney, a former ally of Zada's until he withdrew his investments in one of her projects and then sued her, reported in a deposition that Zada called Perfect 10 "a tax write-off" that he needed to "offset money he made in the market." She also claimed that Zada "needed the loss to represent how small businesses couldn't make money because of piracy on the Internet."[42]

A deposition from one of Zada's targets might not be the best place to find an unbiased account of his dealings, but Chaney did highlight one aspect of Perfect 10's business model: managing the company's relationships with copyright law was one of Zada's major concerns. As a judge in a separate case against Giganews (a Usenet provider) would put it, based on Zada's testimony: "In the lifetime of the company, more than half of Perfect 10's revenues have been generated by litigation. . . . However, all of those revenues were generated by settlements and defaults—Perfect 10 has never obtained a judgement in a contested proceeding in any of its roughly two dozen copyright lawsuits." The strategy proved so fruitful that Zada even purchased copyrights from other creators simply because he "thought they would be helpful in [Perfect 10's] litigation efforts."[43]

Courts were also taking issue with Perfect 10's use of the DMCA. In *Perfect 10 v. Giganews,* the judge concluded that "Perfect 10 has a long, documented history of sending service providers inadequate takedown notices under the DMCA that fail to identify specific infringing material, and then bringing suit for the service providers' failure to respond to deficient DMCA takedown notices."[44] In another case, a global Internet services company called Yandex told a court that it had received hundreds of notices from Perfect 10 regarding a total of 63,756 images found on Yandex servers around the world. Yandex alleged that notices could contain ten pages of written material and hundreds of pages of attachments and that they arrived in batches at a rate more than one per minute.[45]

IMAGE SEARCH

Perfect 10's multiple legal battles forced the courts to grapple with a fundamental question: In what forms can an image exist in the online world? The company's conflict with Yandex provided a partial answer to the question by highlighting instances where an image could be stored on a company's physical servers. This raised jurisdictional problems, as the physical location of servers determined which copyright laws applied, regardless of where the users who uploaded or downloaded the images were located. Thus, Yandex's response to Perfect 10's complaints about the 63,756 images was that 51,485 of them were hosted in servers outside of the United States. Yandex also argued that 50,485 of these were hosted by third-party websites, including 1,464 located on Russian servers.[46]

Despite these jurisdictional complications, this kind of infringement was straightforward compared to the kinds of violations that arose when Perfect 10's images appeared on search engines. This occurred, for example, with Google Images, which launched in 2001. In response to a user's query, Google Images would return a page of image thumbnails. The user could then click on any of the thumbnails, which would display a larger version of the image in a frame within the search results. Clicking on this larger version would take the user directly to the website that hosted the image.

Google Images did not store the images on its servers. Instead, it created frames that displayed the images directly from the host website—allowing the user to see a portion of that site without having to leave the search result page. This meant that a user interested in Perfect 10 could see small- or medium-sized thumbnails of the company's images without leaving Google Images and that the search engine would direct the user to hosting websites regardless of whether those sites had Perfect 10's permission to post the images. Google also created and stored thumbnails of the images on its own servers, though not the full-sized images.

On November 19, 2004, Perfect 10 sued Google on a number of grounds: direct, contributory, and vicarious infringement; circumvention of the DMCA; unfair competition; and several forms of trademark infringement. Perfect 10 estimated that "over 1,000 of [Perfect 10]'s best copyrighted images" were on Google's servers, and it categorized the company's alleged infringement into three groups.[47] First were the thumbnails on the Google search result page, which were "comparable to those available to *perfect10. com* subscribers, and are the same size and clarity as versions" available through Fonestarz.[48] Second were the images a user could see by clicking on a thumbnail: a page displaying the thumbnail and a link to view the full-sized image, along with a note saying, "Image may be scaled down and subject to copyright." When the user clicked on the link, the full-sized image would be displayed in isolation. Third were the cached thumbnails Google stored on its servers, which could be displayed for several months even after the original site deleted the cached image.

One of the points of greatest disagreement between Google and Perfect 10 was over the requirements of a DMCA notice. Perfect 10 claimed to have sent thirty-seven notices, covering more than seven thousand URLs, that identified allegedly infringing images and their source on Perfect 10's site. Google did not immediately take them all down, and Perfect 10 claimed that some remained online four hundred days after the notice. Google responded that the DMCA notices had been defective.[49] The firm claimed that these notices "were unlike any others Google had received, both in volume and incomprehensibility," and that the great majority "failed to identify the specific infringing image and/or which P10 image was infringed."[50] Google had responded by asking Perfect 10 for notices that followed its own guidelines, providing, for example, image-specific URLs instead of the web addresses of sites that displayed dozens or hundreds of images.

SOCCERMANIA.COM

The presiding judge was Howard Matz, who had spent almost thirty years in private practice in New York and Los Angeles before being nominated to the bench in 1997. Matz used an analogy with a fictional website, SoccerMANIA. com (a stand-in for Google), to make sense of this case. Imagine that a soccer enthusiast maintains this website to post personal commentary about recent games, player profiles, and notes on the history of soccer. Imagine that SoccerMANIA.com has a page on the legendary soccer player Pelé, including an image owned by Pelé himself.[51] When visitors arrive at SoccerMANIA. com, they might feel as though the image is *at* SoccerMANIA.com, but that's not necessarily true (figure 18). The website's code could be linking to another site, say SoccerPASSION.com, where the image is stored—and from which the user's computer downloads it directly. (SoccerPASSION was a stand-in for any site where Perfect 10 content is stored.)

In this case, the picture of Pelé never passes through SoccerMANIA's servers. However, since users' browsers display a SoccerMANIA.com URL, users may never realize that the image is hosted at SoccerPASSION.com and downloaded directly into their computers. As Matz put it, "Browsers display the address of the file (here, a webpage) that they are

FIGURE 18. This illustration by my wonderful husband, Vincent Cheng, shows the difference between embedding an image in a website and creating a frame to show an image on another server. At the top, the page SoccerMANIA.com *contains* the image. At the bottom, the page works as a window through which the user can see an image at SoccerPASSION.com. Vincent generously granted me permission to include this illustration and requested the following credit: Vincent Cheng.

currently rendering; they do not in any way indicate the location from which each component element of a webpage (such as an image) originates."[52]

In this instance, Matz continued, a crucial question arises: Which of these two sites, if either, displays the photograph of Pelé when the user visits SoccerMANIA.com? Matz answered this with a legal theory that distinguished two ways of understanding the word "display": one technological, one visual. Technologically speaking, Matz defined "display" as the "act of *serving* content over the web—i.e., physically sending ones and zeroes over the internet to the user's browser."[53] This is the definition Google preferred, since it would mean that SoccerPASSION.com was the one displaying Pelé's image. SoccerMANIA (and therefore Google) would "not risk liability for direct infringement" regardless of whether its linking constitutes fair use. Matz used this reasoning to develop what he called the "server test" of infringement: a website could be liable for infringement if its own servers hosted an image.

Second, there was the user's "purely visual perspective," wherein "display" is "the mere act of *incorporating* content into a webpage that is then pulled up by the browser." In this case, "the act by SoccerMANIA of using an in-line link in its webpage to direct the user's browser to retrieve the Pelé image from SoccerPASSION.com's server each time he navigates to SoccerMANIA.com."[54] This was, of course, Perfect 10's preferred definition. It meant that as the website incorporating the photograph, SoccerMANIA was liable for displaying it. Unlike the server test, the "incorporation test" made a website liable for direct infringement if it incorporated links to copyrighted materials into its HTML code.

Matz saw the server and incorporation tests as "opposite ends of a spectrum," and although both could lead "to extreme or dubious results," he entirely rejected the incorporation test. A person could make a website called "Infringing Content for All!" consisting of thousands of links to copyright-infringing images hosted on other servers. Under the server test, this person would not be liable for direct infringement because the site "does not actually *serve* the images."[55]

Under the incorporation test, anyone who "in-line links to or frames third party content" would be liable for direct infringement, even if they noted the source of the image. In this scenario, SoccerMANIA would be liable even if it included a disclaimer like this: "ATTENTION FBI: We did not take the picture, and it is not served by SoccerMANIA.com. It is probably subject to copyright. We maintain this site to help authorities identify potentially infringing images on the web. The image of Pelé is stored on and served by SoccerPASION.com. Please investigate." Matz concluded that adopting the incorporation test would "cause a tremendous chilling effect on the core functionality of the web" by attacking sites' "capacity to link, a vital feature of the internet that makes it accessible, creative, and valuable."[56]

This way of thinking about full images is essential to social media today. Consider, for example, a recent lawsuit against Instagram. Two photographers, Alexis Hunley and Matthew Brauer, had filed a lawsuit against Instagram for its "embedding feature." You have surely seen this feature, even if you've never noticed: websites sometimes incorporate content from social media platforms such as X or Instagram. Instead of taking a screenshot of the post and putting it on their site, webmasters can embed special code so that the site shows a specific post in the same way that Google Images showed Perfect 10's images. Hunley and Brauer sued Instagram (now part of Meta), arguing that the company should be held liable for vicarious and contributory copyright infringement.

The Ninth Circuit dismissed their case following the *Perfect 10* reasoning very closely. The opinion's summary states that the server test "provides that a copy of a photographic image is not displayed when it is not fixed in a computer's memory" and that the *Perfect 10* decision is not restricted to search engines.[57] But the court noted that the code used to embed an Instagram post into a website did not store or copy a post's image onto the website's servers. Instead, it simply caused the site to create a frame where Instagram's content would appear. The Googlization continues.

The *Field* and *Perfect 10* cases have one very important feature in common: they paved the way for companies to incorporate copyrighted material into their systems without needing creators' permission. *Field* went one step further, by accepting the notion that creators needed to opt *out* of having their text cached. The new default became for Google to cache things. This is an example of what one of my favorite legal scholars, Meg Leta Jones, calls "default dramas": the default settings in platforms that end users may take for granted can hide the intense power struggles among programmers, businesses, users, and law and policy makers.[58]

Perfect 10 allowed thumbnailing for a similar reason: the making of a thumbnail transforms the image from a creative work into a tool for information retrieval. But this case also affirmed an expanded range of acceptable actions, by creating a distinction between the server and incorporation modes of display. Had Matz favored the incorporation test and ruled that any website displaying an image without authorization is committing infringement, then services such as Google Images might have had to stop showing thumbnails in their results pages. We might have entered a new era of the web that recalls the debates following the Scientology lawsuit: Would sites comply with a newfound rule to check every bit of content that goes through their servers, or would they simply modify or cease their services to bypass the requirements?

As with *Field,* the reasoning in *Perfect 10* continues to reverberate today. Its implications were far greater than Perfect 10's specific rights or Google's platforms. Soon after Google's victory, the Electronic Frontier Foundation issued a statement on the opinion: "While it leaves some questions open the bottom line is that the Court upheld important policies of fair use and freedom online and resisted Perfect 10's plea to put copyright owners completely in charge of how and when search engines and other online intermediaries can provide their users with links to images."[59] *Perfect 10* helped to solidify the reasoning that respecting creators' goals for their works is not as important as preserving the platforms that make the web what it is. It's a difficult puzzle: the whole point of copyright is to generate a right to exclude, and yet the free flow of media has become a defining feature of the online environments through which works can circulate.

The makers of generative AI technologies today are building momentum in this same direction, by training their systems on millions of creative works without their makers' consent. The logic so far has been that incorporating a creative work into a training database transforms it into something else. A novel stops being a novel and becomes a string of words, and a photograph of a child stops being a composed snapshot to become an example of how visual elements that can be labeled "child," "face," and "arms" can relate to one another. A few legal battles we'll discuss later are challenging this kind of reasoning. But ask yourself: Could this line of thought be undone without affecting all the other ways in which we experience online media?

CHAPTER NINE

The Biggest Library

WRITING HISTORIES OF THE INTERNET allows one to see how its past folds into its present. Let me, for example, recount five minutes of my research process. When I was writing chapter 5, I remembered that a prominent philosopher named Helen Nissenbaum had written an article in the early 2010s revisiting some of the ideas from her classic book *Privacy in Context*. I Googled two phrases that I remembered from it: "helen nissenbaum" and "privacy online," and there it was: her article "A Contextual Approach to Privacy Online," published in 2011. (You can access it legally via the link in the notes.)[1]

Browsing through Nissenbaum's footnotes, I decided I needed to find a *Wall Street Journal* article titled "What They Know about You." I Googled that title, visited the newspaper's site, and ran into a paywall. I'm sure you've seen these before: soon after you start reading an article, a frame pops up encouraging you to subscribe or purchase access to the work. I have the privilege (thanks to the Sloan Foundation) of being able to purchase whatever I need for my research. But on principle I do not purchase access to websites unless I absolutely need it. I also don't want to stand up, go into the kitchen, and get my phone, which I need to connect to our university's network from home. I'd rather stay seated, and it's distracting to have my phone next to me when I write.

So I went to the Internet Archive (Archive.org), a nonprofit organization founded in 1996 by a librarian named Brewster Kahle, who announced the project as an effort to "construct a digital library." From the first, Kahle acknowledged that the project would raise difficult intellectual property and privacy questions. What if a college student made a "web page that had pictures of her then-current boyfriend" and wanted "to take it down and 'tear it up' "? Or if a candidate for political office wanted to take down something they had written years earlier? Back in the 1990s, before all the developments you've read about, Kahle did not have a clear sense of what was legal in these situations. But he addressed them, at least preemptively, by "allowing authors to exclude their information from the Archive."[2]

Since then, thanks in part to the legal developments we'll discuss in this chapter, the Internet Archive has become enormous. As of September 2024, it makes available more than forty-two million books, almost thirteen million videos, and over a million computer programs. It also maintains the Wayback Machine (https://web.archive.org/), which has been storing copies of websites since the mid-1990s and currently grants access to more than eight hundred billion of them.

At the Wayback Machine, I typed in the URL for the *Wall Street Journal* article and received a chronological list of archived versions of the paper's website, along with a timeline so that I could navigate them. I looked for versions from 2017 because that's when the Wayback Machine archived the website most frequently. I clicked on one of the archived sites and was surprised to run into the paywall again. So I clicked on the earliest archived version, from 2015, and there it was: the text the *Wall Street Journal* had secured behind the paywall.

There are, of course, ways to bypass this paywall—though they certainly violate the DMCA's anticircumvention provisions. For example, it's easy to modify the HTML code of a site. There are browser extensions that allow you to do this quickly, though some of them have been taken down from app stores because of the anticircumvention rules.[3]

But the more important point is that my journey shows a jarring inconsistency: I am not allowed to bypass technical copyright protections for the text of the article, and yet it was perfectly legal for me to access this text by going to a repository of historical websites that has been automatically crawling and storing the web for decades. This mass archiving was made possible by the Googlization of copyright we saw in the last chapter: steadily carving out uses of copyright law that redefined creative works as utilitarian because of what a company is doing to them.

Thus, when I say that the Internet's past is folded into its present, I don't just mean its content. I also mean that the rules that govern our access today embody the infrastructural contradictions that legal developments have enabled over the years. This final chapter follows the project in which everything I have discussed in this book gels together: Google Books.

THE WORLD'S LIBRARY

Google launched Google Print in October 2004 and quickly renamed it Google Books. The plan was to scan magazines and books, make them text-searchable using optical character recognition, and make them widely available (figure 19). Google began by partnering with the New York Public Library and the university libraries at Oxford, Stanford, Harvard, and the University of Michigan. (The company made no distinction between books and magazines, referring to both as "books" early on. I will follow that usage.)

Google's partnership came at just the right time at Harvard, where there was growing momentum to digitize library collections and make them available online. In 2002, a gift from the Flora Family Foundation had enabled the creation of a digital collection called "Women Working, 1870–1930." In 2004, a $5 million private gift from two alumni enabled additional digitization of research materials.[4] Google would now pay to digitize forty thousand books, but the selected books would not follow any thematic or circulation-based rationale: the university would choose them at random.

I was a first semester undergraduate when the partnership was announced, and I remember the talk on campus: Harvard was partnering with this exciting new company to open its massive library holdings to the world. Some of my more senselessly elitist classmates complained that access to the university's resources needed to be earned, but they were swayed by administrators' descriptions of the project as a great act of benevolence and technological prowess. These public relation spins were plentiful. University president Larry Summers claimed that "if this experiment is successful, we have the potential to provide the world's greatest system for dissemination."[5]

For Google, the goal was to enrich search functionality. According to the program's website, "Google's mission is to organize the world's information, but much of that information isn't online yet. Google Print aims to get it there by putting book content where you can find it most easily—right in

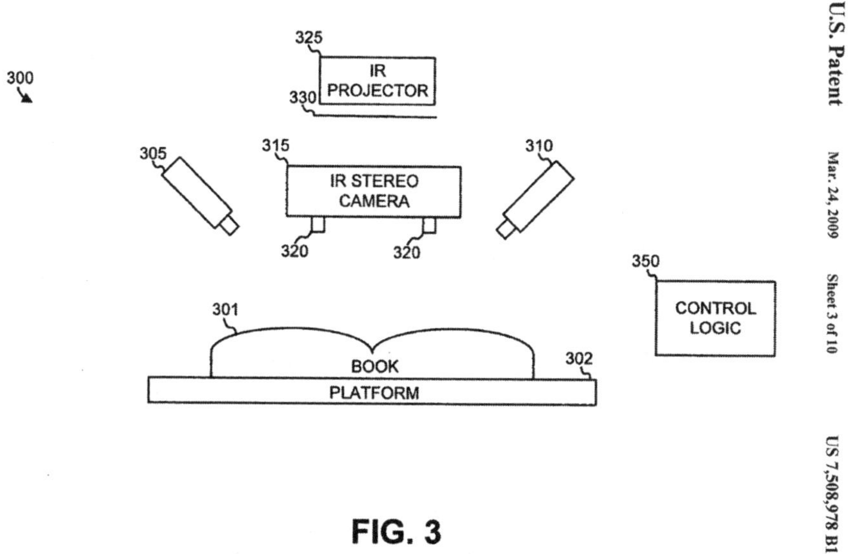

FIG. 3

FIGURE 19. This is a schematic representation of one of Google's patented book scanning technologies. This invention is designed to enable optical character recognition of the book's pages. This image is in the public domain.

your Google search results."[6] A user searching for the phrase "books on ecuador trekking" would see a list of book titles at the top of the search results page. Each title would link to Google's listing for the book, which would include links to the publisher's website and to several online vendors.

A book's listing could show one of three things: a full book (if the work was in the public domain), a full page of the book containing the search term (if the publisher submitted the copyrighted book through Google), or a snippet of the page containing the search term (if Google scanned the book without the publishers' permission through a library partnership). On the left-hand side of the listing, users would find links to key sections of the book, such as the copyright page and table of contents, and a list of websites where they could buy it. The publisher's e-commerce site would be listed at the top, followed by major vendors including Amazon and Barnes and Noble.

Publishers could participate in this project by joining a "Partner Program" that allowed them to submit their books to Google in PDF format, in hard copy, or as a title for one of Google's library partners to digitize. The website for the program in 2005 told publishers that they'd "sell a lot more books if a lot more people knew about them" and that they should "think of

Google Book Search as a free worldwide sales and marketing system."[7] Google later advised publishers that they could increase traffic to their own websites by buying their own ads on other websites or by hosting advertisements of their own.[8]

And this is why Google Books quickly became controversial: publishers could volunteer their books for digitization and searchability, but Google was going to digitize them and store them on its servers whether publishers agreed to it or not (figure 20). I was too busy learning how to write an essay and do English-language math to worry about copyrights, but I remember that among my classmates were two groups of folks who were appalled by indiscriminate library scanning: libertarians, who thought of property rights as one of the most sacred legal frameworks; and aspiring authors, who saw this as a threat to their future professions.

The discontent I saw on campus was brewing on the national level. Faced with complaints from publishers and authors, Google announced in August 2005 that it would stop scanning library books until November. This would allow each publisher to make a list of titles it did not want Google to scan. But the trouble remained: having a book scanned would be the *default* option, not a copyright owner's choice. Patricia Schroeder, president of the Association of American Publishers, commented that "Google's procedure shifts the responsibility for preventing infringement to the copyright owner rather than the user, turning every principle of copyright law on its ear."[9]

On September 20, 2005, the Authors Guild—the largest trade group for authors in the country—filed a class action lawsuit against Google to stop the Library Program.[10] The complaint alleged that Google was committing infringement because it was violating (among other things) authors' exclusive right to reproduce, distribute, and display their works. It requested an injunction prohibiting Google from continuing its unauthorized reproduction through the Library Program and a judgment declaring Google's actions unlawful. The Association of American Publishers filed a similar suit, and the two were eventually merged, so let's keep our focus on the Authors Guild.

The Authors Guild's suit named three authors who volunteered to present their complaints of copyright infringement: nonfiction author Herbert Mitgang, children's and young-adult author Betty Miles, and poet and essayist Daniel Hoffman. Mitgang charged Google with infringing his

FIGURE 20. In 2008, books from the University of Michigan's Harlan Hatcher Graduate Library were taken off the shelves while Google digitized them. This image was uploaded to Wikimedia Commons by mollyali, and I modified it only to increase its resolution. It is licensed under the Creative Commons Attribution 2.0 Generic License (https://creativecommons.org/licenses/by/2.0/deed.en).

copyright on a biography of Abraham Lincoln; Miles, on a work of fiction; and Hoffman, on a book of poems and a scholarly work on John Milton.

All the works these authors cited in their complaint had been published in the late 1960s and the early 1970s, which meant they were protected by the Copyright Act of 1909 and not the 1976 act. Congress passed the 1909 act at a time when cinema and recorded music were unprecedented new technologies, and it sided with the new industries growing up around these technologies by extending copyright protection to works in the new media. The Authors Guild may have been preparing to argue that the new situation was a modern analog of that earlier time.

The Authors Guild's complaint prompted Google to revise how it presented its digitization project. In December 2004, before the complaint, the main website for the project (http://print.google.com/googleprint/library.html) described its library partnerships this way: "Our library partnerships vastly increase the amount of material that we're able to digitize and bring online. Authors and publishers will benefit from the increased visibility of current books, and over time, Google users will even be able to find out of print and rare books that were previously unavailable except from a library. Books still in copyright will be searchable, but users will only be able to view bibliographic information and a few small text snippets—similar to the experience of flipping through a book in a bookstore."[11]

Compare this description, with its evocation of flipping through a book at a bookstore, with what the same URL said in September 2005: "This project's aim is simple: make it easier to find relevant books. We hope to guide more users to books—specifically books they might not be able to find any other way—all while carefully respecting authors' and publishers' copyrights. Our ultimate goal is to work with publishers and libraries to create a comprehensive, searchable, virtual card catalog of all books in all languages that helps users discover new books and publishers find new readers."[12]

No longer the virtual equivalent of browsing through a bookstore, the Library Project is now a "virtual card catalog." The difference between the two metaphors is legally significant: it is much more difficult to argue that Google is infringing on publishers' and authors' copyrights if a court believes the company is creating digital card catalogs, not digital bookshelves.

The catalog metaphor directs attention away from the entire books Google had stored in its servers and focuses on users' access to a library-style listing: an organized and searchable database that shows users where they can find potentially copyrighted material. The "cards"—the individual book listings—included all the information that a traditional card catalog would contain and a very small snippet of copyrighted material for every search term a user entered. This had the potential to defuse complaints that the company's page displays constituted copyright infringement and to narrow the legal issue down to whether it was infringement for Google to scan the books and store them in its servers.

On this narrower question, Google had an advantage thanks to *Kelly v. Arriba,* the case that in 2003 established the use of thumbnails in search engines as fair use. The Ninth Circuit didn't mind that Arriba was using artists' thumbnails for a commercial purpose (running its profitable search engine). What mattered to the court was that the company's use of the art was "more incidental and less exploitative in nature than more traditional types of commercial use." According to the court: "Arriba was neither using

Kelly's images to directly promote its web site nor trying to profit by selling Kelly's images. Instead, Kelly's images were among thousands of images in Arriba's search engine database. Because the use of Kelly's images was not highly exploitative, the commercial nature of the use weighs only slightly against a finding of fair use."[13]

The Ninth Circuit supplemented this with an explanation that Arriba's use was transformative. The thumbnails were much smaller and had a lower resolution than Kelly's images, and they served "an entirely different function." They were no longer "artistic works intended to inform and engage the viewer in an aesthetic experience" but "a tool to help index and improve access to images on the internet."[14]

This reasoning was Google's most powerful legal weapon. As the lawyer Jonathan Band put it, "Everything the Ninth Circuit stated with respect to Arriba applies with equal force to the Print Library Project."[15] *Arriba* also aligned Google's interest with that of powerful Internet advocacy groups, particularly the Electronic Frontier Foundation, which tended to favor removing barriers to the free online circulation of information.

DON'T BE EVIL

By 2006, it was becoming clear that Google's famous slogan, "Don't be evil," was failing as a moral compass. On February 20, the cover of *Time* featured Google CEO Eric Schmidt and cofounders Larry Page and Sergey Brin—all wearing black, with their arms crossed, sporting smug smiles—next to the question, "CAN WE TRUST GOOGLE WITH OUR SECRETS?"[16] Adi Ignatius wrote the cover story, titled "In Search of the Real Google," in which he asked what it meant for a company to be good when it was already worth $100 billion and making unilateral decisions about what data to keep and store.

Ignatius's point was that Brin and Page's goal of making "nearly all information accessible to everyone all the time" was difficult to categorize as good.[17] He recalled that Google had been slammed for a software feature that results in the company's storing users' personal data for up to a month." And just a week before featuring the chief executives on its cover, *Time* reported that Google had launched a website in China that gave access only to content sanctioned by the Chinese government. In Ignatius's eyes, actions like these killed the slogan: "Doing a totalitarian government's bidding in blocking the truth in order to make a few extra bucks is practically the definition of evil."[18]

Ignatius's report also highlighted how the secrecy of Google's strategy belied the appearance of transparency that Page and Bryn tried to cultivate. He noted that Brin and Page squirmed when asked if Google has a master plan. "So what's the plan?" he asked. "World domination? Keep throwing money at everything and see what works?" All the reporter knew was that the list of Google's one hundred top priorities was confidential and that "Google will keep looking for new ways to organize and search for information," making money primarily through advertising.[19] Everything else was a guess.

The problem is that Google's "Don't be evil" slogan had become "Do what the law allows you to do." Even Vint Cerf, the Internet founding father who became Google's "chief Internet evangelist," admitted this. In 2006, commenting on search censorship in China, he noted that "there's a subtext to 'Don't be evil,' and that is 'Don't be illegal.' "[20] But following the law is not the same as avoiding evil—and doing the right thing can require illegal behavior. Legality and goodness can be completely separate, yet Google was taking them as one and the same.

This confusion was happening with library scanning. *Kelly v. Arriba* meant that Google had every reason to think that scanning library books without permission was legal as long as the scans became part of a search engine. Libraries' and readers' support of the project added a secondary line of evidence for the perceived goodness of its actions. Many authors and publishers thought that unauthorized digitization was not right—even if it was legal. So Google would need to win two battles: the legal one initiated by the Authors Guild and a public relations one meant to preserve its reputation.

Google launched aggressive public relations campaigns that celebrated the company as a force for good. This was especially important for the Library Project, which was already attracting negative press and threatened to become a lengthy legal battle. The campaign to celebrate the project was not subtle. Several Google employees took to the company's Books blog (booksearch.blogspot.com) to share personal stories of how Google Books allowed them to find information that they would have never been able to access without it.[21]

Company representatives attended professional meetings for librarians to spread their gospel and gather testimony about how noble their efforts were. One of the most hyperbolic statements Google published on its blog was attributed to Laura Moody, a librarian at the Gulf Shores Public Library in Alabama, who wrote in an email, "I admire a company that is so in line with the premise of library ideals. The dissemination of knowledge, especially the scanning in of the university collections and making them available to all in the WORLD, is truly noble. You are not just working at a job. You will enable, in a strange way, the remaining doors of ignorance to be blasted away. You will allow those people who are still isolated from the rest of the world to experience the light of knowledge. Remain true to your vision. Go home at the end of the day with a sigh of well-earned satisfaction. Some people talk about changing the world. You are actually doing it."[22] Pam Saenger, an associate product marketing manager at Google, wrote in a blog post that Moody's email had made her team "think more deeply about how similar Google's mission is to a librarian's, and to explore ways we can work more closely with librarians in the quest to connect people and information."[23]

Libraries were generating the support the company needed to continue arguing that its digitization effort was, in fact, a force for good. On August 9, 2006, for instance, a University of Michigan librarian named John Wilkin wrote a blog post welcoming the University of California to the Library Project in which he celebrated the way more and more libraries were "making a significant contribution to the dissemination of knowledge worldwide."[24] In their own press release, University of California administrators lauded Google's digitization as an unmitigated good, whose benefits ranged from furthering the university's mission to preventing damage from natural disasters such as Hurricane Katrina, which had recently devasted libraries in Louisiana and Mississippi.[25]

This expanding landscape of support was making publishers and authors look like the evil ones, fighting a losing battle against technological change and the free flow of information. Consider, for example, a blog post titled "Why Google Is Right" by Susan Crawford, a prominent legal scholar who would later become Barack Obama's special assistant for science, technology, and innovation policy. "The authors who are suing," she wrote, "are claiming that they'd like to license their works for online searching themselves—and they're free to do that. They can simply ask Google not to include them in Google's pool. Their claim is that that's too much of a burden. Phooey. Google's inclusion of their books in the searchable pool can only help these authors, not hurt them."[26]

A chorus of academics and practitioners echoed Crawford's thoughts. Lawrence Lessig said in 2006 that "Google Print could be the most important contribution to the spread of knowledge since Jefferson dreamed of national libraries." He thought that the Copyright Office was such a broken agency that it was impossible to ask for permission from authors (even though Google had proposed a Publisher Program meant to address this). He also noted that if it was legal for Google to index the Internet itself, it should be allowed to digitize and index books. "The 'authors' ' claims," he concluded, "if true, mean Google itself is illegal. Common sense, or better, commons sense, revolts at the idea. And so too should you."[27]

So library-based scanning continued and even intensified, despite authors' and publishers' complaints. In 2006 and 2007, the libraries at the University of Wisconsin–Madison, Complutense University of Madrid, the National Library of Catalonia, and the University of Texas at Austin all joined the project, as did many others.[28] Google maintained a list of quotes from prominent scholars and practitioners noting how important and obviously right it was to support the effort. Cindy Cohn, from the EFF, said it would be a shame if the debate became "a binary choice between authors and publishers on the one hand and Google on the other." She noted a third interest, "those of 'us' who are trying to find the right thing to read."[29]

COMPETITION AND PRIVACY

In October 2008, Google proposed a settlement with the Authors Guild. The full proposal is a stunning document: three hundred pages of text so dense that early reports, even from the EFF, included a disclaimer saying that analysts were still digesting it.[30] The settlement would have perpetuated Google's right to keep scanning and transformed the company into a bookseller. Google would have the right to sell institutional subscriptions to an electronic database of books and to sell online access to individual books. Google would also have the right to "sell advertising on pages from Books" and to a list of "other uses" that I'll explain in a moment. In exchange, Google would pay 63 percent of the revenue these uses generated back to copyright owners and disburse at least $45 million in cash payments to owners of books digitized before May 5, 2009.[31]

The crux of the settlement was Google's proposal to establish and manage a nonprofit "Book Rights Registry," which would serve as a copyright clearinghouse and royalty allocator. It would collect contact information about copyright owners, document any requests they made regarding the use of their works, and coordinate payments. The company proposed to pay $34.5 million to create the registry, part of which would defray the costs of notifying copyright owners and administering the settlement until the registry was operational. The registry would be copyright owners' representative before Google; it would be managed by a board comprising at least four authors and four publishers.

This settlement proposal was wildly unpopular. One of the major questions it raised was whether Google's ability to digitize and sell books would harm authorship and publishing. William Cavanaugh, a deputy assistant attorney general in the Justice Department's Antitrust Division, filed a brief claiming that Google was attempting to use the Authors Guild's class action suit "to implement forward-looking business arrangements that go far beyond the dispute before the Court."[32] The arrangement would have granted Google legal rights that conflicted with copyright owners' well-established right to "control whether and how to exploit their works," giving anticompetitive advantages only to Google. It would make Google "the only competitor in the digital marketplace with the rights to distribute and otherwise exploit a vast array of works in multiple formats."[33]

Lawyers and legal scholars across the country expressed similar concerns, often emphasizing the proposal's handling of orphan books—works that are protected by copyright but whose authors are unknown. Google would have had the exclusive right to digitize these works. New York Law School professor James Grimmelmann told the *New York Times* that the settlement "gives Google a free pass for infringement for selling all these books"; even the publishers themselves were to be excluded from Google's monopoly on orphan books.[34]

The Antitrust Division's involvement is not surprising. The Justice Department been a major actor in computing and telecommunications since the Kingsbury Commitment of 1913, in which AT&T agreed to divest itself from Western Union and all other non-telephone businesses. Historically, the division had enforced antitrust law through warnings and consent decrees—agreements by companies to modify their behavior to avoid or end a lawsuit. And Google was already in the DOJ's crosshairs. Earlier in 2008, it had warned Google and Yahoo! that they would face a lawsuit if they implemented a plan to have Yahoo! replace a large portion of its own search page advertisements with ads sold by Google. This move, the DOJ warned, would make the companies collaborators and reduce competition in the search engine industry.[35] (Coincidentally, while I was writing this chapter, on January 24, 2023, the Department of Justice sued Google once again, this time for monopolizing digital advertisement technologies.)[36]

What surprised me when I first started researching this lawsuit was that the settlement's privacy implications became a major hurdle. A seemingly harmless article by Motoko Rich published in the January 4, 2009, issue of the *New York Times* raised major red flags among online privacy advocates.[37] The article was titled, "Google Hopes to Open a Trove of Little-Seen Books," and it was, for the most part, a puff piece celebrating how Google Books was giving folks access to out-of-copyright books that were otherwise hard or impossible to find.

Perhaps inadvertently, Rich reported a major violation of user privacy. To make her point that users were using Google's digital holdings, she wrote that Daniel Clancy, the director of engineering for the project, was "monitoring search queries" when a user's query for "concrete fountain molds" caught his attention. According to Rich, "The search turned up a digital version of an obscure 1910 book, and the user had spent four hours perusing 350 pages of it."[38] This showed that Google's staff had the ability to see what books individual users searched for, which ones they read, and perhaps most shockingly, how long they spent reading them.

Google did not release a "Google Books Privacy Policy" until September 3, 2009, the very last day before the court-mandated filing date for objections to the settlement. The timing made people question whether Google was trying to sweep some major problems under the rug. The policy was short, a page and a half long, and meant as an addendum to Google's main policy. It noted that Google "will receive information similar to what it would receive for a web search: the query term, IP address, browser type and language, a time stamp, and one or more cookies that may uniquely identify your browser."[39] A closing section specific to the settlement agreement promised that users' information would not be shared with third parties unless there was a legal requirement to do so and that the company would limit credit card companies' access to users' purchases.

Bad timing aside, what stands out from this proposal is that it would allow Google to access detailed search and reading data about each user. It did not hide the fact that it knew what users were searching for: in August 2009, *New York Times* columnist Maureen Dowd visited Google headquarters and was surprised to see the company flaunting this knowledge.[40] She reported that "there is a vaguely ominous Big Brother wall in the lobby of the headquarters here that scrolls real-time Google searches—porn queries are edited out—from people around the world. You could probably see your own name if you stayed long enough. In one minute of watching, I saw the Washington association where my sister works, the Delaware beach town where my brother vacations, some Dave Matthews lyrics, calories Panera, females feet, soaps in depth and Douglas Mangum, whoever he is."[41]

If Google's September 3 privacy policy acknowledged that the data it collected on books resembled search data, what could it know about its users? Would someone like Daniel Clancy be able to see that a person is looking for books about queerness to explore their own gender in secret? And regardless of the *content* of the searches: If Google was monetizing users' reading habits through ad placement, should it be constrained in the information it "knows" about us?

Concerns such as these motivated the EFF, the ACLU, and several authors and scholars to band together as the "Privacy Authors and Publishers" and file a brief opposing Google's proposed settlement. They emphasized that Google would be able to merge reading habit data with everything else it knew about individual users indefinitely and that recent reports on how easy it is to de-anonymize data made this problem potentially permanent.[42]

People sometimes need privacy to access information. Have you ever chosen to browse the Internet on "Incognito" mode so that your computer does not record your browsing history? Searching for pornography is perhaps the most popular use of that browsing mode, but there are many other circumstances in which people might need anonymity. Think about a teenager growing up in a homophobic family who wants to know more about what the term "LGBTQIA" means. Or perhaps someone who needs information about suicide prevention and doesn't want their roommates to know.

The Privacy Authors and Publishers knew that protecting anonymity had long been legally recognized as essential to ensuring the circulation of ideas. In 1953, Supreme Court Justice William O. Douglas wrote: "Once the government can demand of a publisher the names of the purchasers of his publications, the free press as we know it disappears. Then the spectre of a government agent will look over the shoulder of everyone who reads. The purchase of a book or pamphlet today may result in a subpoena tomorrow. Fear of criticism goes with every person into the bookstall. The subtle, imponderable pressures of the orthodox lay hold. Some will fear to read what is unpopular, what the powers-that-be dislike. When the light of publicity may reach any student, any teacher, inquiry will be discouraged."[43]

This kind of reasoning had extended well beyond books. In the mid-1990s, the Court noted that cable subscribers would hesitate to file written requests for sexual programming out of "fear for their reputations should the operator . . . disclose the list of those who wish to watch the 'patently offensive' channel."[44] There is even an instance when customers boycotted a bookstore because they thought the store had revealed a customer's purchase history to a grand jury.[45]

The EFF rallied Privacy Authors and Publishers and librarians' professional organizations in calling for stronger privacy protections. Their joint public statements included a quote from novelist Jonathan Lethem that summarized their common ground: "Now is the moment to make sure that Google Book Search is as private as the world of physical books. If future readers know that they are leaving a digital trail for others to follow, they may shy away from important but eccentric intellectual journeys."[46]

Denny Chin, whom President Bill Clinton had appointed to the District Court of the Southern District of New York in 1994, was overseeing the case. Chin was born in Hong Kong in 1954 and had graduated from Princeton University and Fordham Law School. In the 1980s, he had made a name for himself prosecuting fraud, corruption, and organized crime. At the Second District, Chin had continued to handle criminal cases involving high-profile defendants such as casino magnate Sheldon Adelson and Ponzi schemer Bernard Madoff.

On September 24, 2009, Chin delayed holding a hearing on the fairness and reasonableness of the Google–Authors Guild settlement proposal in light of all the issues the parties and others had raised. He was not swayed by any one argument but was struck by the breadth of communities from which these arguments came. The battle over the settlement would continue for many months, until Judge Chin ultimately rejected an amended agreement in March 2011.

In his rejection, Chin cited concerns over privacy, copyright, antitrust, and even international law, but the core of his reasoning was something that had been a problem from the very beginning: the settlement would "transfer to Google certain rights in exchange for future and ongoing arrangements" and would "release Google (and others) from liability from certain *future* acts."[47] Chin also rejected the class action status of the case, on the grounds that the interests of all parties affected by the suit were not adequately represented. Before the case moved on to a full trial, however, Google had one more tool at its disposal: calling for summary judgment.

By then, courts had made it clear that the inclusion of a creative work in a database was a transformative use. Cases such as the ones on Google Images and on Arriba Soft's and Google's search engines had all provided plenty of justification for Chin to rule that Google's scanning, storing, and limited display of books constituted fair use. In addition, a similar lawsuit launched by the Authors Guild against HathiTrust (a spinoff of Google Books hosted partly by the University of California) seemed to be headed toward a victory for the repository.

Based on this growing tradition of legal reasoning, Chin found that "Google Books is also transformative in the sense that it has transformed book text into data for purposes of substantive research, including data mining and text mining in new areas, thereby opening up new fields of research. Words in books are being used in a way they have not been used before. Google Books has created something new in the use of book text—the frequency of words and trends in their usage provide substantive information."[48]

The Second Circuit Court of Appeals affirmed this reasoning in 2015, ruling that Google's digitizing was "highly transformative" and thus non-infringing fair use. It didn't matter that Google had a commercial intention, because the digitized books "do not provide a significant market substitute for the protected aspects of the originals."[49] Google was even allowed to provide libraries with digital copies of the scanned books without becoming a contributory infringer. The Supreme Court declined to hear an appeal the next year, and Google Books continued to grow.

The web has become an archive of our human experience. Our books, news, research, and art are all in there—as are our personal expressions of joy, grief, rage, and everything else. But the trouble is that this archive is not a passive space because it is continually crawled and processed, sometimes by the technical systems on which it runs. The extent and intensity at which this is happening are unprecedented in the history of technology. Think of what it took for Google or Microsoft to incorporate ChatGPT into their search engines. When I search for anything now, I receive an unsolicited summary of what the most popular websites for my search term have to say. This advice is not so much an expression of collective wisdom or consensus as it is what results when a generative AI system processes search results after having digested the archive that we have all been building.

By the time this book is published, this might all have changed yet again. In 2023, the publisher of the *New York Times* found that certain prompts would cause ChatGPT to output the text of some of their more popular articles, word by word. At the end of the year, the publisher filed a lawsuit against OpenAI (maker of ChatGPT) and its largest backer, Microsoft. The Times's complaint alleged that the AI technologies "were built by copying and using *millions* of The Times' copyrighted news articles, in-depth investigations, opinion pieces, reviews, how-to guides, and more."[50] The suit is still pending, but I cannot help to think about the thick line that connects this brand-new conflict to 2 Live Crew's historic victories at the Supreme Court, the Googlization of copyright law, and the recalibrations of fair use that are likely to stem from Justice Sotomayor's opinion in *Warhol v. Goldsmith*. I therefore leave you with an open question: How much could online copyright law change without radically destabilizing the online world to which we've grown accustomed?

Epilogue

WHATEVER YOUR ETHICAL STANCE ON copyrights, they are fundamental to our online experiences. Copyrights have *made* the online world in the sense that they condition, and sometimes even determine, the range and modes of engagement that we can have with creative works in that world. This occurs at all levels and across all devices, even if we don't realize it.

The history of online copyright is ultimately about how we incorporate creativity into our daily lives. Our experiences online unfold at the very turbulent intersection between the most powerful corporations in the world and our own desires for media consumption. Consider how nerve-racking it must be for an aspiring artist or musician to build a social media presence, knowing that they are giving the biggest platforms in the world permission to do what they want with their art and that it takes just one person to make one of their works commercially unviable. If you browse through porn, think about the copyright notices that show up as screens warning of lawsuits and FBI penalties, either on the websites themselves or as small see-through texts embedded into the videos and pictures. Think about the millions of creative works necessary for generative AI technologies to transform how we think, learn, and see the world—and take a second during your next mindless newsfeed scrolling to identify every copyrighted work that zooms past you at the swipe of a finger.

I hope you've come to appreciate how our relationships with online media shift alongside transformations in US copyright law. Change can happen slowly or overnight. Some is happening right now, and we are still waiting to see where it takes us. Let me conclude with three areas of transformation I see in our near future.

UPDATING THE DMCA

The Digital Millennium Copyright Act is almost twenty-five years old, and we engage with the online world very differently now than when the law was drafted in the 1990s. It is certainly due for an update. As pressure to do so mounts, I keep reflecting on the tensions that develop between legislation and the everyday realities of how people use networked devices. Efforts to prohibit ways of engaging with online content have had unexpected consequences that both push that mode of engagement underground and create new possibilities for misusing copyright law.

If you think back to the Scientology and *Playboy* chapters, you'll recall that the DMCA resolved an issue that courts were having to address case by case: determining whether an online service provider was liable for its users' copyright infringements. In the twentieth century, copyright owners were able to sue service providers directly, and providers were not required to take down infringing content. The safe harbor provisions of the DMCA created, at least in theory, a way for copyright owners and service providers to resolve these kinds of conflicts without resorting to the courts. The notice and takedown process is the central component of this system, but it proved very easy to abuse, and it facilitated new forms of online censorship.

One idea that has been floating around the legal news cycles is creating a "notice and *stay* down" system. North Carolina senator Thom Tillis introduced this idea in 2020 as part of a bill called the Digital Copyright Act of 2021. If it had passed, whenever a copyright owner submitted a notice to an online service provider, the provider would have had to ensure that all "complete or near complete" copies of the work in question were deleted entirely from its servers. The DMCA's current system targets individual instances of a work, but this update would target *all* instances (including, for example, a person sending a video privately through WhatsApp).[1]

The privacy and censorship considerations of this proposal are very serious. Recall how fragile user privacy was when Viacom was investigating copyright infringements on YouTube, and think about the extent to which targeting *individual* instances of a work has led to a global pattern of censorship that we are only starting to understand. I worry that a stay-down system would strengthen the misuse of copyright as a tool for censorship, and not just in the United States. Before even considering a stay-down system, I would want to see a lot more research on the transformation of US copyright into a global force for online censorship.

FAIR AND TRANSFORMATIVE USES

The legacy of *Campbell v. Acuff Rose Music* lives on. This was the Supreme Court opinion involving the Miami-based rap group 2 Live Crew, in which the Court ruled that their parody of "Oh, Pretty Woman" constituted fair use. Its ruling defined a transformative use as one that "adds something new, with a further purpose or different character." Since that ruling, transformative use has become one of the central considerations in high-profile copyright lawsuits.

For example, when I started writing this book, a high-stakes battle between Google and Oracle was the headliner of copyright news. It involved about eleven thousand lines of copyrighted code for something called an application program interface. An API, as folks call this program, allows two computer programs to communicate with each other (in this case, phone apps and operating systems). Oracle owned about eleven thousand lines of code, and Google copied them exactly into its Android operating system. Google claimed that this was fair use, and in a 6–2 opinion in 2021, the Supreme Court agreed. Justice Stephen Breyer, who wrote the opinion, found that Google's use of the code was transformative because Google "copied only what was needed to allow programmers to work in a different computing environment without discarding a portion of a familiar programming language."[2] He cited *Campbell* to justify calling this a transformative use. Google got its way yet again, this time having arguably plagiarized eleven thousand lines of code.

In the aftermath of the *Warhol* opinion, what we mean by fair use seems to be up for grabs. Justice Elena Kagan disagreed with the majority opinion. She thought that the Court was wrong to declare "that Andy Warhol's eye-popping silkscreen of Prince—a work based on but dramatically altering an existing photograph—is (in copyright lingo) not 'transformative.' " Warhol had "effected a transformation" by adding "something new and important" that Conde Nast and its readers could appreciate. Indeed, she insisted that the work was transformative because of "differences in meaning that arose from replacing a realistic—and indeed humanistic—depiction of the performer with an unnatural, disembodied, masklike one."[3]

Instead of clarifying what uses are *not* transformative, *Warhol* pushes us to dive into the gray area between fair and unfair uses and try to find some structure in it. The stakes are only getting higher, as copyright holders challenge using their works in training data sets for generative AI. Will the Googlization of copyright lead courts to allow this, or will Sotomayor's restrictions on what it means for a use to be "transformative" change the course of the conversation?

AI AND COPYRIGHT

This is the big one. It ties together everything we know about copyright and asks us to reconsider our most fundamental assumptions about it. But there are two big problems that I find particularly interesting. The first is how we should allocate copyrights over works created by or with the assistance of an AI system.

There are several schools of thought on this issue. One approach proposes that works created by AI cannot be copyrighted and belong in the public domain. Others argue that AI output should depend on its creator's intent and input. But what would happen, for example, if I trained an AI system to write text using only the scholarly works I have written? Why shouldn't I be able to claim a copyright over whatever text it generates? What if I have already used AI to generate this book? Or if I use AI to write a fantastic book that becomes a bestseller? Do I have no control over it?

That seems extreme, but that would be the implication if we use the case of Naruto (the monkey that took a selfie) to determine who owns AI-created works. Back then, the court didn't rule that David Slater—the photographer who set the cameras in the jungle—owned the copyrights. It simply ruled that the photograph could not be copyrighted because it was not a human creation. We need new precedent. The Naruto case is best known for affirming that nonhuman animals can't hold copyrights; it's time to figure out what happens when we take animals out of the equation.

Another approach proposes that AI-generated works should be treated like works-for-hire. In this framework, using an AI system to create work is analogous to hiring someone do it. Whoever used the AI, like whoever hired a person, owns the copyrights. But this doesn't solve the problem. In the scenario involving my own works, someone else who runs my system would own copyrights over text that was created based only on texts I have written. Does this mean I have to keep my personally trained AI a trade secret? Restrict access to it so that only people who have agreed not to claim copyrights can use it? What if someone took my system and modified it for their own purposes? Would I have any claim over the texts *they* made?

This brings me to the second big problem: AI is using works without their creators' permission. In the scenario involving my own works, everything was mine. But all the generative AI systems that are circulating on the web right now rely on other people's works without their consent. These systems are crawling through the online world, collecting and classifying images and texts, and generating new works based on them (figure 21). This is partly enabled by the doctrine of transformative use: we could argue, quite convincingly, that the incorporation of a work of art into an AI system is transformative because the work is being used for a new purpose—as part of a training database instead of as a creative work. We would mirror the arguments that Google used to defend its cache and image search results page.

Technology can't solve this problem. Some efforts are underway to develop tools that prevent crawlers from processing an image into training databases. This would be analogous to adding code into a website to stop Google's bot from storing it in its cache. But recall how easy it was for researchers to bypass audio watermarks and crack Adobe's e-publication software. Remember all the pornography we can stream and download even though companies enforce their digital rights very aggressively. Technological fixes are Band-Aids on an open wound—simply not enough.

Legal efforts to address the AI issue are already underway. For example, in July 2023, a judge dismissed a lawsuit by a group of artists against three generative artificial intelligence companies, asking the companies to provide more facts about what the artists claimed was copyright infringement. But courts alone can't solve this, either. Think about how the transformative use doctrine developed for decades: courts systematically sided with technological change over the desires of artists whose work was circulating without their permission.

I don't know what will happen next. I always tell my students that my job is not to predict the future or tell them how the world *should* be. It is to help them grapple with, and I hope participate in, whatever change comes our

 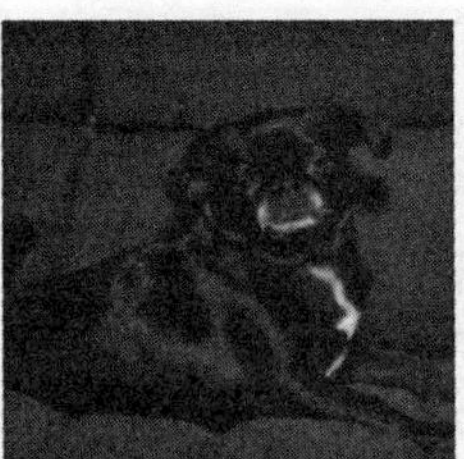  

FIGURE 21. Our dog Luna stares at me while I make pictures of her using DALL-E 2. The other images were produced by Dall-E2 when I used the prompt "a black dog with a white tuft of hair on the center of her chest, the dog is a little pudgy and a mix between pug and chihuahua, on a dark blue couch."

way according to their own values. This is true for you, too. I can't predict how these problems will be solved because, as this book has shown, a very influential lawsuit can emerge out of nowhere and shake up everything we knew about online copyright. Only you can decide what you want for the works you create. But I can assure you that no matter what happens next for creative works online, it will involve the things you've learned from this book.

Thank you for reading. And one more thing: if you put a PDF of this book onto the web, please don't make the link public.

INTRODUCTION

1. See Gerardo Con Díaz, *Software Rights: How Patent Law Transformed Software Development in America* (New Haven: Yale University Press, 2019).

2. See Safiya Umoja Noble, *Algorithms of Oppression: How Search Engines Reinforce Racism* (New York: NYU Press, 2018); and Virginia Eubanks, *Automating Inequality: How High-Tech Tools Profile, Police, and Punish the Poor* (New York: St. Martin's, 2018).

3. This field of study is large, but I am very inspired by the work of B. Zorina Khan, Petra Moser, Jonathan Barnett, Sean Bottomley, and three of my own mentors, Daniel Kevles, Mario Biagioli, and Naomi Lamoreaux. Nowadays, I keep returning to B. Zorina Khan, *Inventing Ideas: Patents, Prizes, and the Knowledge Economy* (New York: Oxford University Press, 2020).

4. My two favorite works on this topic are D. Graham Burnett's *Trying Leviathan: The Nineteenth-Century New York Court Case That Put the Whale on Trial and Challenged the Order of Nature* (Princeton, NJ: Princeton University Press, 2007); and Jill Lepore's essay "On Evidence: Proving *Frye* as a Matter of Law, Science, and History," *Yale Law Journal* 124, no. 4 (2015): 882–1345. A useful introductory overview to the legal studies of information technology is Nicolas P. Suzor, *Lawless: The Secret Rules That Govern Our Digital Lives* (Cambridge: Cambridge University Press, 2019).

5. U.S. Const., art. I, § 8, cl. 8.

6. The four books I use to introduce graduate students to internet infrastructures are Janet Abbate, *Inventing the Internet* (Cambridge, MA: MIT Press, 2000); Ya-Wen Lei, *The Contentious Public Sphere: Law, Media, and Authoritarian Rule in China* (Princeton, NJ: Princeton University Press, 2017); Laura DeNardis, *The Internet in Everything: Freedom and Security in a World with No Off Switch* (New

Haven: Yale University Press, 2019); and Nicole Starosielski, *The Undersea Network* (Durham, NC: Duke University Press, 2015).

PART ONE. SHARING

1. Alexandra C. Bell, "New Law Bans Film Piracy," *Harvard Crimson,* November 29, 2004, https://www.thecrimson.com/article/2004/11/29/new-law-bans-film-piracy-massachusetts/.

2. Bell, "New Law Bans Film Piracy," archived at Wayback Machine, citing a capture dated June 4, 2015, https://web.archive.org/web/20150604013122/https://www.thecrimson.com/article/2004/11/29/new-law-bans-film-piracy-massachusetts/.

3. Ishtiaq Ahmad (@ishtiaq_ahmad9), "I will remove leaked onlyfans content under dmca," Fiverr sale listing, https://www.fiverr.com/ishtiaq_ahmad9/leaked-onlyfans-content-onlyfans-content-protection-leaked-patreon-content-dmca.

4. Ishtiaq Ahmad (@ishtiaq_ahmad9), "Dmca Attorney," Fiverr seller profile, https://www.fiverr.com/ishtiaq_ahmad9.

5. Ahmad, "Dmca Attorney."

6. @maarya1, screenshot image of links requested to be removed and completed requests in review on Ahmad, "I will remove leaked onlyfans content under dmca," ca. late 2023, https://fiverr-res.cloudinary.com/image/upload/c_limit,f_auto,q_auto,t_smartwm,w_500/v1/attachments/delivery/asset/5d169d88d51cf31c163ad04f144b310a-1690159139/IMG_20230724_053404.jpg.

7. @maddycoles, "Responds very fast. Reasonable pricing/flexibility, esp compared to others offering the same services on this site," review on Ahmad Fiverr seller profile, "I will remove leaked onlyfans content under dmca," ca. late 2023, https://www.fiverr.com/ishtiaq_ahmad9.

8. @cleosummers, "Im happy that he did what i asked in way of reporting the leaked content but i am not sure it has been removed. It appears to still be live," review on Ahmad Fiverr seller profile, ca. 2022–23, https://www.fiverr.com/ishtiaq_ahmad9.

9. Shreya Tewari, "Over Thirty Thousand DMCA Notices Reveal an Organized Attempt to Abuse Copyright Law," Lumen, April 22, 2022, https://www.lumen-database.org/blog_entries/over-thirty-thousand-dmca-notices-reveal-an-organized-attempt-to-abuse-copyright-law.

10. Tewari, "Over Thirty Thousand DMCA Notices."

CHAPTER ONE. SCIENTOLOGY ONLINE

1. This story is based on Dennis L. Erlich, "This morning at 7:30 am my doorbell was repeatedly rung by an [*sic*] man in a suit with papers. I didn't open and when he didn't go away I called 911 . . .," InFormer Ministry, ca. February 1997, archived at Wayback Machine, capture dated June 10, 1997, https://web.archive.org/web/19970610012829/http://eff.org/pub/Censorship/Scientology_cases/erlich_experience.article.

2. Erlich, "This morning at 7:30 am."

3. You can find the video of this encounter at https://web.archive.org/web/20240531215418/https://www.youtube.com/watch?v=dauM5JhFYXY.

4. Erlich, "This morning at 7:30 am."

5. See Church of Scientology Int'l v. Fishman, No. CV 91-6426 (C.D. Cal. 1991).

6. This general discussion about the Church of Scientology is grounded on Ann Brill and Ashley Packard, "Silencing Scientology's Critics on the Internet: A Mission Impossible?," *Communications and the Law* 19, no. 4 (1997): 1–24.

7. Brill and Packard, "Silencing Scientology's Critics," 4.

8. United States v. Ballard, 322 U.S. 78 (1944), cited in Brill and Packard, "Silencing Scientology's Critics," 7.

9. Alison Frankel, "Making Law, Making Enemies," *American Lawyer,* March 1996, 79.

10. Articles of Incorporation for the Religious Technology Center, available online at https://en.wikipedia.org/wiki/File:RTC-Incorporation.pdf.

11. See Elizabeth A. Rowe, "Trade Secret Litigation and Free Speech: Is It Time to Restrain the Plaintiffs?," *Boston College Law Review* 50, no. 5 (2009): 1425–1455.

12. See Jemillah Bowman Williams, "Diversity as a Trade Secret," *Georgetown Law Review* 107, no. 6 (2019): 1685–1732.

13. For trade secret protection, see Bridge Publ'ns, Inc. v. Vien, 827 F. Supp. 629 (S.D. Cal. 1993); and Religious Tech. Ctr. v. Scott, 869 F.2d 1306 (9th Cir. 1989). For spiritual injury, see Religious Tech. Ctr. v. Wollersheim, 796 F.2d 1076 (9th Cir. 1986), *cert. denied,* 479 U.S. 1103, 107 S. Ct. 1336, 94 L. Ed. 2d 187 (1987). See also Brill and Packard, "Silencing Scientology's Critics," 4–6.

14. Brill and Packard, "Silencing Scientology's Critics," 3.

15. New York Times Co. v. Sullivan, 376 U.S. 254, 270 (1964).

16. Andrew Ventimiglia, *Copyrighting God: Ownership of the Sacred in American Religion* (Cambridge: Cambridge University Press, 2019).

17. Kevin Driscoll, *The Modem World: A Prehistory of Social Media* (New Haven: Yale University Press, 2022), 3.

18. Scott Goehring, quoted in Wendy M. Grossman, "alt.scientology.war," *Wired,* December 1, 1995, https://www.wired.com/1995/12/alt-scientology-war/.

19. Grossman, "alt.scientology.war"; Wendy M. Grossman, *Net.Wars* (New York: New York University Press, 1997).

20. "Scientology Cases," Electronic Frontier Foundation, capture dated February 25, 1997, archived at Wayback Machine, https://web.archive.org/web/1997022 5050656/http://www.eff.org/pub/Censorship/Scientology_cases/.

21. Brill and Packard, "Silencing Scientology's Critics," 10.

22. Brill and Packard, "Silencing Scientology's Critics," 10.

23. Dennis Erlich, quoted in Grossman, "alt.scientology.war."

24. Dennis Erlich, quoted in Alan Abrahamson and Nicholas Riccardi, "Scientologists Seize Disks for Lawsuit," *Los Angeles Times,* February 14, 1995; reposted by an193903@anon.penet.fi, "La times raid 2."

25. Brill and Packard, "Silencing Scientology's Critics," 10.

26. Verified Complaint for Injunctive Relief and Damages, Religious Tech. Ctr. v. Netcom On-Line Comm. Servs., Inc., No. C-95-20091 RMW (N.D. Cal.

February 7, 1995), quoted in "Rtc v netcom et al complaint," BBS post to alt.religion.scientology, February 27, 1995, reposted by Electronic Frontier Foundation, archived at Wayback Machine, citing a capture dated August 15, 2000, https://web.archive.org/web/20000815232308/http://www.eff.org/pub/Legal/Cases/Scientology_cases/cos_020895.complaint.

27. For more on the early history of BBSs, see Driscoll, *Modem World*.

28. Thomas Klemesrud, "HELENA KOBRIN CONTACT," BBS post to alt.religion.scientology, December 30, 1994, 13:41:04, https://groups.google.com/g/alt.religion.scientology/c/vgtOssuCinA/m/tqWh_86_cwoJ; Helena Kobrin, "Infringements through SUPPORT.com," December 30, 1994, 19:12:59 PST, quoted in Dennis L. Erlich, "infringements through support.com," December 30, 1994, 20:46:15, https://groups.google.com/g/alt.religion.scientology/c/ElgUPnNNrBg/m/l1DcJKsA3eAJ.

29. Thomas Klemesrud, "Infringements through 2/2," BBS post to alt.religion.scientology, December 30, 1994, 23:24:47, https://groups.google.com/g/alt.religion.scientology/c/mxe8As7pRPM/m/FUg1_MMZyhUJ.

30. Klemesrud, "Infringements through 2/2."

31. Verified Complaint for Injunctive Relief and Damages, RTC v. Netcom.

32. Reporter's Transcript of Proceedings, RTC v. Netcom, No. C-95-20091-RMW (N.D. Cal. February 21, 1995), edited for readability by Ron Newman, reposted by Electronic Frontier Foundation, archived at Wayback Machine, citing a capture dated August 15, 2000, https://web.archive.org/web/20000815232314/http://www.eff.org/pub/Legal/Cases/Scientology_cases/court_hearing_022195.transcript.

33. Order for Temporary Restraining Order, Order to Show Cause Re: Preliminary Injunction, and for Order of Impoundment, RTC v. Netcom, No. C-95-20091 RMW (N.D. Cal. February 10, 1995), quoted in Lee Holzinger, "Temporary Restraining Order against Erlich," BBS post to alt.religion.scientology, February 16, 1995, reposted by Electronic Frontier Foundation, archived at Wayback Machine, citing a capture dated August 15, 2000, https://web.archive.org/web/20000815234806/http://www.eff.org/pub/Legal/Cases/Scientology_cases/erlich_tro_021095.order.

34. Declaration of Rick Francis in Support of Defendant Netcom's Opposition to Plaintiffs' Request for Injunctive Relief, RTC v. Netcom, No. C-95-20091 RMW (N.D. Cal. February 21, 1995), reposted by Electronic Frontier Foundation, archived at Wayback Machine, citing a capture dated August 15, 2000, https://web.archive.org/web/20000815234845/http://www.eff.org/pub/Legal/Cases/Scientology_cases/francis_021795.declaration.

35. Declaration of Francis.

36. Reporter's Transcript of Proceedings.

37. Andrew H. Wilson, quoted in Reporter's Transcript of Proceedings.

38. Reporter's Transcript of Proceedings.

39. Tarleton Gillespie, *Wired Shut: Copyright and the Shape of Digital Culture* (Cambridge, MA: MIT Press, 2007).

40. Ronald M. Whyte, quoted in Reporter's Transcript of Proceedings.

41. Thomas M. Small, quoted in Reporter's Transcript of Proceedings.

42. Dennis Erlich, quoted in Reporter's Transcript of Proceedings.

43. Erlich, quoted in Reporter's Transcript of Proceedings.

44. Whyte, quoted in Reporter's Transcript of Proceedings.

45. On this subject, see Michael Hanne and Robert Weisberg, eds., *Narrative and Metaphor in the Law* (New York: Cambridge University Press, 2018).

46. Gerardo Con Díaz, *Software Rights: How Patent Law Transformed Software Development in America* (New Haven: Yale University Press, 2019).

47. See, e.g., Annette N. Markham and Katrin Tiidenberg, eds., *Metaphors of Internet: Ways of Being in the Age of Ubiquity* (New York: Peter Lang, 2020).

48. Con Díaz, *Software Rights*. See also "Networks for the People: Clinton-Gore and the Information Superhighway," available online at https://clintonwhitehouse4. archives.gov/WH/New/Starbright/Network.html.

49. Randolf J. Rice, quoted in Reporter's Transcript of Proceedings.

50. Memorandum of Points and Authorities of Defendant Netcom On-Line Communication Services, Inc. in Opposition to Plaintiffs' Request for Injunctive Relief, RTC v. Netcom, No. C-95-20091 RMW (N.D. Cal. February 21, 1995), archived at Wayback Machine, citing a capture dated June 10, 1997, https://web. archive.org/web/19970610013253/http://eff.org/pub/Censorship/Scientology_ cases/netcom_opposition_021795.memo.

51. Andrew H. Wilson, quoted in Reporter's Transcript of Proceedings.

52. Rice, quoted in Reporter's Transcript of Proceedings.

53. Richard A. Horning, quoted in Reporter's Transcript of Proceedings.

54. City of Los Angeles, Municipal Code § 41.01.1 (1959), quoted in Smith v. California, 361 U.S. 147, 148 (1959).

55. Memorandum of Points and Authorities of Defendant Netcom On-Line Communication Services, Inc. in Opposition to Plaintiff's Request for Injunctive Relief.

56. Amended Temporary Restraining Order, RTC v. Netcom, No. C-95-20091 RMW (N.D. Cal. February 23, 1995), quoted in "RTC v. Netcom—Amend TRO," BBS post to alt.religion.scientology, February 27, 1995, reposted by Electronic Frontier Foundation, archived at Wayback Machine, citing a capture dated August 15, 2000, https://web.archive.org/web/20000815234547/http://www.eff.org/pub/ Legal/Cases/Scientology_cases/amend_erlich_tro_022395.order.

57. Helena K. Kobrin, fax and letter to Ronald M. Whyte, February 27, 1995; reposted by Dennis L. Erlich, "Kobrin jumps on an er 1/2," post on alt.religion. scientology, February 27, 1995, 19:41:46 PST; archived at Wayback Machine, citing a capture dated March 2, 1997, https://web.archive.org/web/1997030 2163829/http://www.eff.org/pub/Censorship/Scientology_cases/kobrin_ whyte_022795.letter.

58. Declaration of Dennis Erlich in Support of Opposition to Motion for OSC Re Contempt, RTC v. Netcom, No. C-95-20091 RMW (N.D. Cal. March 17, 1995), archived at Wayback Machine, citing a capture dated March 2, 1997, https://web. archive.org/web/19970302164236/http://www.eff.org/pub/Censorship/ Scientology_cases/erlich_031495.declaration.

59. Kobrin to Whyte.

60. Dennis L. Erlich to Ronald Whyte, February 27, 1995; archived at Wayback
Machine, citing a capture dated March 2, 1997, https://web.archive.org/
web/19970302163756/http://www.eff.org/pub/Censorship/Scientology_cases/
erlich_whyte_022795.letter.

61. Kobrin to Whyte.

62. Grossman, "alt.scientology.war."

63. Brian A. Carlson, "Balancing the Digital Scales of Copyright Law," *SMU Law
Review* 50, no. 3 (1997): 825–866.

64. Electronic Frontier Foundation, "EFF Opposes Scientology Censorship and
Attacks on System Operators," *EFFector Online* 8, no. 2 (1995), https://www.eff.
org/effector/8/2.

65. Electronic Frontier Foundation, "EFF Opposes Scientology Censorship."

66. Electronic Frontier Foundation, "EFF Opposes Scientology Censorship."

67. The Electronic Frontier Foundation has archived a transcript of her testimony:
https://web.archive.org/web/19970302163542/http://www.eff.org/pub/
Censorship/Scientology_cases/munsey_021795.declaration.

68. Dennis L. Erlich, "Rosa v. Dennis Erlich 1" and "INFORMER'S INTENT," e-mail to
Ron Newman reposted to alt.religion.scientology, March 29, 1995, 23:26:54 PST,
and March 25, 1995, 8:17:15 PST, archived at Wayback Machine, citing a capture
dated March 2, 1997, https://web.archive.org/web/19970302164352/http://www.
eff.org/pub/Censorship/Scientology_cases/erlich_munsey_032995.reply.

69. Declaration of Warren McShane, in Declarations of Warren McShane and Rosa
Erlich Munsey in Support of Plaintiffs' [*sic*] Reply in of Order to Show Cause Re
Civil Contempt, RTC v. Netcom, No. C-95-20091 RMW (N.D. Cal. March 17,
1995), reposted by Electronic Frontier Foundation, archived at Wayback
Machine, citing a capture dated August 15, 2000, https://web.archive.org/
web/20000815235040/http://www.eff.org/pub/Legal/Cases/Scientology_cases/
mcshane_031695.declaration.

70. Dennis Erlich, post to alt.religion.scientology, March 10, 1995, quoted in
Declaration of McShane.

71. Dennis Erlich, February 24, 1995, quoted in Declaration of McShane.

72. Anonymous, quoted in Declaration of McShane.

73. Dennis Erlich, response to Anonymous, quoted in Declaration of McShane.

74. Mark Walsh, "MoFo Takes Copyright Case of Scientology Church Critic,"
Recorder, March 15, 1995, archived by Wayback Machine, citing a capture dated
March 2, 1997, https://web.archive.org/web/19970302164256/http://www.eff.
org/pub/Censorship/Scientology_cases/walsh_erlich_031595.article.

75. Walsh, "MoFo Takes Copyright Case."

76. Dennis L. Erlich, "JUDGE PUTS HOLD ON CASE," BBS post to alt.religion.scientology,
March 16, 1995, 5:41:45, https://groups.google.com/g/alt.religion.scientology/c/
wSQCM89TDZc/m/yEVWSlJvKPgJ.

77. Order Granting in Part and Denying in Part Plaintiffs' Application for a
Preliminary Injunction and Defendant Erlich's Motion to Dissolve the TRO;
Denying Plaintiffs' Application to Expand the TRO; Denying Plaintiffs' Motion

for Contempt; Granting Erlich's Motion to Vacate the Writ of Seizure; and
Denying Plaintiffs' Request for Sanctions against Erlich's Counsel, RTC v.
Netcom, No. C-95-20091 RMW (N.D. Cal. September 22, 1995), quoted in
Diane Richardson, "Decision in RTC (Church of Scientology) v. Netcom, et al.,
923 F. Supp. 1231 (1995)," post to alt.religion.scientology, September 27, 1995,
archived at Wayback Machine, citing a capture dated June 10, 1997, https://web.
archive.org/web/19970610013355/http://eff.org/pub/Censorship/Scientology_
cases/whyte_cos_v_erlich_092295.ruling.

78. See Amended Notice of Motion and Motion to Dismiss for Failure to State a
Claim upon Which Relief Can Be Granted (FRCP 12(b)(6)) or, in the Alternative,
Summary Judgment, RTC v. Netcom (N.D. Cal. April 14, 1995), reposted by
Electronic Frontier Foundation, archived at Wayback Machine, citing a capture
dated August 15, 2000, https://web.archive.org/web/20000815235105/http://
www.eff.org/pub/Legal/Cases/Scientology_cases/netcom_dismiss_031395.
motion. See also Renewed Motion for Preliminary Injunction against
Defendants Netcom On-Line Communications Services, Inc. and Tom
Klemesrud, March 6, 1995, reposted by Electronic Frontier Foundation, archived
at Wayback Machine, citing a capture dated March 2, 1997, https://web.archive.
org/web/19970302164121/http://www.eff.org/pub/Censorship/Scientology_
cases/renew_cos_injunction_030895.motion.

79. Memorandum of Points and Authorities of Defendant Netcom On-Line
Communication Services, Inc. in Support of Motion to Dismiss or, in the
Alternative, Summary Judgment, RTC v. Netcom, No. C-95-20091 RMW
(N.D. Cal. April 14, 1995), reposted by Electronic Frontier Foundation, archived
at Wayback Machine, citing a capture dated August 15, 2000, https://web.
archive.org/web/20000815235111/http://www.eff.org/pub/Legal/Cases/
Scientology_cases/netcom_dismiss_support_031395.memo.

80. Memorandum of Points and Authorities of Defendant Netcom in Support of
Motion to Dismiss or, in the Alternative, Summary Judgment.

81. Declaration of Rick Francis in Support of Netcom's Motion to Dismiss, or in the
Alternative, for Summary Judgment, RTC v. Netcom, C-95-20091 RMW (N.D.
Cal. April 14, 1995), quoted in Memorandum of Points and Authorities of
Defendant Netcom in Support of Motion to Dismiss or, in the Alternative,
Summary Judgment.

82. Memorandum of Points and Authorities of Defendant Netcom in Support of
Motion to Dismiss or, in the Alternative, Summary Judgment.

83. This paragraph is based on Fonovisa, Inc. v. Cherry Auction, Inc., 847 F. Supp.
1492 (E.D. Cal. 1994). The Ninth Circuit opinion is Fonovisa, Inc. v. Cherry
Auction, Inc., 76 F.3d 259 (9th Cir. 1996).

84. Fonovisa v. Cherry (1994), at 1496.

85. See Kevin Driscoll, "Hobbyists Inter-Networking and the Popular Internet
Imaginary: Forgotten Histories of Networked Personal Computing, 1978–1998"
(PhD diss., University of Southern California, 2014).

86. Order Denying Defendant Netcom's Motion for Summary Judgment; Denying
Defendant Klemesrud's Motion for Judgment on the Pleadings; and Denying

Plaintiffs' Motion for Preliminary Injunction against Netcom and Klemesrud, RTC v. Netcom, No. C-95-20091 RMW (N.D. Cal. November 21, 1995), reposted by Electronic Frontier Foundation, archived at Wayback Machine, citing a capture dated August 15, 2000, https://web.archive.org/web/20000815235205/http://www.eff.org/pub/Legal/Cases/Scientology_cases/whyte_netcom_112195.order.

87. Jeff Kosseff, *The Twenty-Six Words That Created the Internet* (Ithaca, NY: Cornell University Press, 2019).

88. Sarah T. Roberts, *Behind the Screen: Content Moderation in the Shadows of Social Media* (New Haven: Yale University Press, 2019).

CHAPTER TWO. THE *PLAYBOY* SUITS

1. My favorite book on this topic is Alexander Monea, *The Digital Closet: How the Internet Became Straight* (Cambridge, MA: MIT Press, 2022).

2. Faraz Ahmed, M. Z. Shafiq, and Alex X. Liu, "The Internet Is for Porn: Measurement and Analysis of Online Adult Traffic" (paper presented at 2106 IEEE 36th International Conference on Distributed Computing Systems [ICDCS], Nara, Japan), 88–97.

3. United States v. Playboy Entm't Grp., Inc., 529 U.S. 803 (2000).

4. United States v. Playboy, 813.

5. Nadine Strossen, *Defending Pornography: Free Speech, Sex, and the Fight for Women's Rights* (1995; new ed., New York: NYU Press, 2000).

6. P. Gregory Springer, "Tandy's Magnificent Concession: The Model 1000 Is IBM PC Compatible and a Good Choice for Home Use with Its Deskmate Software," *Infoworld,* June 3, 1985, 72; archived by Google Books, https://books.google.com/books?id=8C4EAAAAMBAJ&pg=PA72.

7. Russ Hardenburgh, "Rusty n Edie's from Beginning to the Beginning," BBS post, March 15, 1990, available online at http://www.textfiles.com/bbs/ad-art.txt.

8. Hardenburgh, "Rusty n Edie's."

9. Hardenburgh, "Rusty n Edie's."

10. Hardenburgh, "Rusty n Edie's."

11. Hardenburgh, "Rusty n Edie's."

12. Hardenburgh, "Rusty n Edie's."

13. Hardenburgh, "Rusty n Edie's."

14. Hardenburgh, "Rusty n Edie's."

15. Hardenburgh, "Rusty n Edie's"; "Here's the Story of How Rusty and Edie Got Started in the 'BBS Business,' " IPTIA Consulting, https://www.ipingthereforeiam.com/bbs/articles/Pioneering_BBSes/Rusty_n_Edies/.

16. Hardenburgh, "Rusty n Edie's."

17. "Rusty & Edie's BBS Seized by the FBI," *Boardwatch Magazine,* April 1993, 30, available online at https://archive.org/details/Boardwatch1993-04/page/n29/mode/2up.

18. For more on the Silicon Valley myth, see Joy Lisi Rankin, *A People's History of Computing in the United States* (Cambridge, MA: Harvard University Press, 2018).

19. "Rusty & Edie's BBS Seized by the FBI."

20. "Rusty & Edie's BBS Seized by the FBI," 30–31.

21. Con Díaz, *Software Rights*.

22. "Rusty & Edie's BBS Seized by the FBI," 30.

23. "Rusty & Edie's BBS Seized by the FBI," 30.

24. "Rusty & Edie's BBS Seized by the FBI," 31.

25. Bruce Sterling, *The Hacker Crackdown: Law and Disorder on the Electronic Frontier* (New York: Bantam Books, 1992); Steve Jackson Games, Inc. v. United States Secret Service, 816 F. Supp. 432 (W.D. Tex. 1993).

26. "Rusty & Edie's BBS Seized by the FBI," 31.

27. "Playboy Interview: Steven Jobs," *Playboy*, February 1985, 49–70, 174–184.

28. Elizabeth Fraterrigo, *Playboy and the Making of the Good Life in Modern America* (New York: Oxford University Press, 2009).

29. "Playboy Joins the Computer Revolution," April 2, 1987, available online at https://www.upi.com/Archives/1987/04/02/Playboy-joins-computer-revolution/1116544338000/.

30. Eileen Kent, quoted in "Playboy Joins the Computer Revolution."

31. Paul McGraw, quoted in "Playboy Joins the Computer Revolution."

32. Dylan Mulvin, *Proxies: The Cultural Work of Standing In* (Cambridge, MA: MIT Press, 2021).

33. Copyright Act of 1976, Pub. L. No. 94-553, 90 Stat. 2541 (1976).

34. Copyright Act of 1976, 2542.

35. Playboy Enterprises, Inc. v. Frena, 839 F. Supp. 1552, 1556 (M.D. Fla. 1993).

36. Con Díaz, *Software Rights*.

37. *Final Report of the National Commission on New Technological Uses of Copyrighted Works* (Washington, DC: Library of Congress, 1979), 84.

38. Sega Enterprises, Ltd. v. MAPHIA, 857 F.Supp. 679 (N.D. Cal. 1994).

39. Sega v. MAPHIA, 686.

40. Playboy Enterprises, Inc. v. Russ Hardenburgh, Inc., 982 F.Supp. 509 (1997).

41. Playboy v. Hardenburgh, 512.

42. Playboy v. Hardenburgh, 503, 512.

43. Susan Decker and Christopher Yasiejko, "Porn Purveyors' Use of Copyright Lawsuits Has Judges Seeing Red," Bloomberg, August 5, 2019, https://www.bloomberg.com/news/articles/2019-08-05/porn-purveyors-use-of-copyright-lawsuits-has-judges-seeing-red.

44. "Received a Strike 3 Holdings Subpeona Letter," The Russell Firm, https://www.russellfirmip.com/lp/strike-3-holdings-lawsuits-new/; "Have You Received a Summons or Subpoena from Strike 3 Holdings LLC?," Antonelli Law, https://www.antonelli-law.com/bittorrent-isp-subpoena/strike-3-holdings-llc/.

CHAPTER THREE. THE NAPSTER HYDRA

1. "What Is Napster?," Napster, archived at Wayback Machine, citing a capture dated February 29, 2000, https://web.archive.org/web/20000229185025/http://napster.com/whatisnapster.html.

2. "Home," Napster, archived at Wayback Machine, citing a capture dated March 1, 2000, https://web.archive.org/web/20000301102915/http://www.napster.com.

3. "Home."

4. Sony Corp. of America v. Universal City Studios, Inc., 464 U.S. 417 (1984).

5. Kevin Driscoll, *The Modem World: A Prehistory of Social Media* (New Haven: Yale University Press, 2022).

6. "WIPO Copyright Treaty (WCT)," World Intellectual Property Organization, https://www.wipo.int/treaties/en/ip/wct/.

7. World Performances and Phonograms Treaty, *adopted* December 20, 1996, arts. 10, 14.

8. Directive 2001/29/EC of the European Parliament and of the Council of 22 May 2001 on the Harmonisation of Certain Aspects of Copyright and Related Rights in the Information Society, 2001 O.J. (L 167), art. 6(1), http://data.europa.eu/eli/dir/2001/29/oj.

9. 17 U.S.C. § 512.

10. Linton Weeks, "Turning Up the Power: Recording Industry Lobbyist Hilary Rosen, in Tune with Washington," *Washington Post*, July 29, 1997.

11. Weeks, "Turning Up the Power."

12. Brian Hiatt, "Metallica, Zeppelin, Billy Joel Honored for 10 Million-Plus Sales: Their Records Were among 62 Honored with Industry's Newest Award for Units Shipped," *MTV*, March 16, 1999, archived at Wayback Machine, citing a captured dated August 16, 2016, http://www.mtv.com/news/512893/metallica-zeppelin-billy-joel-honored-for-10-million-plus-sales/.

13. Weeks, "Turning Up the Power."

14. Weeks, "Turning Up the Power."

15. For more on file formats, see Jonathan Sterne, *MP3: The Meaning of a Format* (Durham, NC: Duke University Press, 2012); and Paratii, "A Brief History of P2P Content Distribution, in 10 Major Steps," *Medium*, October 25, 2017, https://medium.com/paratii/a-brief-history-of-p2p-content-distribution-in-10-major-steps-6d6733d25122.

16. Iowa666SiC, "Lars Ulrich," YouTube video clip of 2000 MTV Awards, June 18, 2007, 0:27 to 0:33, https://www.youtube.com/watch?v=_qoZ3gBActg.

17. Iowa666SiC, "Lars Ulrich."

18. David Kirkpatrick, "With a Little Help from His Friends," *Vanity Fair*, October 2010, archived at Wayback Machine, citing a capture dated January 21, 2015, 2024, https://web.archive.org/web/20150121060241/http://www.vanityfair.com/culture/features/2010/10/sean-parker-201010.

19. "Home," Napster, archived at Wayback Machine, citing a capture dated October 8, 1999, https://web.archive.org/web/19991008215720/http://napster.com/.

20. Robert Lemos, "Napster Plays Dodgeball with Music Biz: Start-Up Offers Chat with a Twist—Access to Music Files on Others' Computers; But Is It Legal?," ZD Net News, August 13, 1999, archived by Wayback Machine, citing a capture dated August 15, 2000, http://web.archive.org/web/20000815204211/http://www.zdnet.com/zdnn/stories/news/0,4586,2314760,00.html.

21. "Napster Software User Agreement," Napster, quoted in Jennifer Sullivan, "Napster: Music Is for Sharing," *Wired News,* November 1, 1999, archived by Wayback Machine, citing a capture dated March 3, 2000, http://web.archive.org/web/20000303053901/http://www.wired.com/news/print/0,1294,32151,00.html.

22. Jim Griffin, quoted in Sullivan, "Napster."

23. MTV News Staff, "Digital Nation: Napster, MP3 Matchmaker: New Program Links Music Fans with One Another So They Can Trade Files," MTV, November 2, 1999, http://www.mtv.com/news/519542/digital-nation-napster-mp3-match maker/ (accessed October 2021; URL no longer active).

24. Lanny Udey, quoted in Lisa Guernsey, "MP3-Trading Service Can Clog Networks on College Campuses," *New York Times,* January 20, 2000.

25. Amy Harmon, "Potent Software Escalates Music Industry's Jitters," *New York Times,* March 7, 2000.

26. Paul Heltzel, "Students Have Few Qualms about Online Music Piracy," *New York Times,* May 2, 2000.

27. Heltzel, "Students Have Few Qualms."

28. Heltzel, "Students Have Few Qualms."

29. Don Tapscott, quoted in Heltzel, "Students Have Few Qualms."

30. Mark Pastin, quoted in Heltzel, "Students Have Few Qualms."

31. "Homepage," Soundbyting, archived by Wayback Machine, citing a capture dated March 2, 2000, http://web.archive.org/web/20000302152954/http://www.soundbyting.com/.

32. "SoundByting: Top 10 Myths," Soundbyting, archived by Wayback Machine, citing a capture dated March 2, 2000, http://web.archive.org/web/20000302191448/http://www.soundbyting.com/html/top_10_myths/myths_index.html.

33. Kristen Philipkoski, "University Snoops for MP3s, page 1," *Wired,* November 13, 1999, archived by Wayback Machine, citing a capture dated March 1, 2000, http://web.archive.org/web/20000301235058/http://wired.com/news/technology/0,1282,32478,00.html.

34. Gary Kendrick, quoted in Kristen Philipkoski, "University Snoops for MP3s, page 1," *Wired,* November 13, 1999, archived by Wayback Machine, citing a capture dated March 3, 2000, http://web.archive.org/web/20000303141414/http://www.wired.com/news/technology/0,1282,32478,00.html.

35. Kendrick, quoted in Philipkoski, "University Snoops for MP3s, page 1," Wayback Machine capture dated March 3, 2000.

36. Jennifer Sullivan, "MP3 Pirate Gets Probation," *Wired,* November 24, 1999, https://www.wired.com/1999/11/mp3-pirate-gets-probation/.

37. Kristen Philipkoski, "University Snoops for MP3s, page 2," *Wired,* November 13, 1999, archived by Wayback Machine, citing a capture dated March 3, 2000, http://web.archive.org/web/20000303165936/http://www.wired.com/news/technology/0,1282,32478-2,00.html.

38. Lydia Pelliccia, quoted in "RIAA Suing Upstart Startup," *Wired News,* November 15, 1999, archived at Wayback Machine, citing a capture dated March 2, 2000,

http://web.archive.org/web/20000302195631/http://www.wired.com/news/
print/0,1294,32559,00.html.

39. Cary Sherman, quoted in Jack McCarthy, "Music Trade Group Sues MP3
Start-Up," *ComputerWorld,* December 9, 1999, archived at Wayback Machine,
citing a capture dated March 1, 2000, http://web.archive.org/web/20000
301144027/http://www.computerworld.com/home/news.nsf/
all/9912094recordsue.

40. Eileen Richardson, quoted in Guernsey, "MP3-Trading Service."

41. Hank Barry, quoted in Matt Richtel, "Napster Has a New Interim Chief and Gets
a $15 Million Investment," *New York Times,* May 23, 2000.

42. "Disclaimer: What You Should Know about Napster and MP3," Napster, archived
at Wayback Machine, citing a capture dated November 27, 1999, http://web.
archive.org/web/19991127123703/http://www.napster.com/disclaimer.html.

43. "Terms of Use," Napster, archived at Wayback Machine, citing a capture dated
February 29, 2000, https://web.archive.org/web/20000229161718/http://www.
napster.com/terms.html.

44. "Terms of Use."

45. Ron White and Michael White, *MP3 Underground: The Inside Guide to MP#
Music, Mapster, Realjukebox, Musicmatch, and Hidden Internet Songs* (2000; repr.,
Indianapolis: Que Pub, 2001).

46. White and White, *MP3 Underground,* inner cover.

47. White and White, *MP3 Underground,* inner cover.

48. White and White, *MP3 Underground,* 110–113.

49. White and White, *MP3 Underground,* 113.

50. David Weekly, "The Napster Protocol," capture dated May 20, 2000, https://web.
archive.org/web/20000520100035/http://david.weekly.org/code/napster.php3.

51. Paul Festa, "Copyright Defendant Napster on the Other Side in Property
Dispute," *CNET News,* January 26, 2000, capture dated May 16, 2000, https://
web.archive.org/web/20000516094702/http://news.cnet.com/news/0-1005-
200-1531752.html.

52. *Downloaded,* directed by Alex Winter (VH1, 2013).

53. 35 U.S.C. § 271.

54. Sony v. Universal.

55. Recording Indus. Ass'n of America v. Diamond Multimedia Sys., Inc., 180 F.3d
1072 (9th Cir. 1999).

56. A&M Records, Inc. v. Napster, Inc., 239 F.3d 1004 (9th Cir. 2001).

57. Robin D. Gross, "9th Circuit Napster Ruling Requires P2P Developers Ensure
No One Misuses Their Systems: Supreme Court's 'Betamax' Defense to
Secondary Liability Narrowed, Appeals Court Requires Judge to Rewrite
Software to Prevent Infringement," Electronic Frontier Foundation, February
26, 2001, archived at Wayback Machine, citing a capture dated December 31,
2002, https://web.archive.org/web/20021231061317/http://www.eff.org/Legal/
Cases/Napster_cases/20010226_rgross_nap_essay.html.

58. A&M Records v. Napster, 1021.

59. MGM Studios, Inc. v. Grokster, Ltd., 545 U.S. 913, 919 (2005).

60. "Help and Frequently Asked Questions," Grokster, archived at Wayback Machine, citing a capture dated August 3, 2001, https://web.archive.org/web/200108 03093858/http://grokster.com/helpfaq.html#What%20is%20Grokster.

61. "Welcome to Grokster!," Grokster, archived at Wayback Machine, citing a capture dated July 22, 2001, https://web.archive.org/web/20010722035636/http://grokster.com/.

62. "Welcome to Grokster!"

63. "About Us," Grokster, archived at Wayback Machine, citing a capture dated August 3, 2001, https://web.archive.org/web/20010803093029/http://grokster.com/aboutus.html.

64. "Artist Services," Grokster, archived at Wayback Machine, citing a capture dated November 1, 2001, https://web.archive.org/web/20011101034238/http://www.grokster.com/contentproviders.html. See also "Press Information," Grokster, archived at Wayback Machine, citing a capture dated November 10, 2001, https://web.archive.org/web/20011110183827/http://www.grokster.com/press.html.

65. "Terms of Service," Grokster, archived at Wayback Machine, citing a capture dated November 10, 2001, https://web.archive.org/web/20011110180800/http://www.grokster.com/policies.html.

66. "Terms of Service," Grokster, archived at Wayback Machine, citing a capture dated June 8, 2001, https://web.archive.org/web/20010608102522/http://www.grokster.com/policies.html.

67. Email exchange between Tyler Nilson and support@grokster.com, started February 10, 2002, in joint Appendix 3, MGM v. Grokster, 939.

68. Email exchange between Nilson and support@grokster.com.

69. Email from Shadowdancer1130@aol.com to support@grokster.com, February 28, 2002, in Joint Appendix 3, MGM v. Grokster, 947.

70. "RIAA v. The People: Five Years Later," Electronic Frontier Foundation, September 30, 2008, archived at Wayback Machine, citing a capture dated March 6, 2023, https://web.archive.org/web/20230306080154/https://www.eff.org/wp/riaa-v-people-five-years-later.

71. "RIAA v. The People: Two Years Later," Electronic Frontier Foundation, archived at Wayback Machine, citing a capture dated February 6, 2002, https://web.archive.org/web/20060206093653/https://www.eff.org/IP/P2P/RIAAatTWO_FINAL.pdf; 17 U.S.C. § 512(f).

72. "RIAA v. The People: Two Years Later," 5–6.

73. "Sample RIAA Complaint," Electronic Frontier Foundation, archived at Wayback Machine, citing a capture dated November 1, 2008, https://web.archive.org/web/20081101113654/http://w2.eff.org/IP/P2P/sample_riaa_complaint.pdf.

74. *Kim Dotcom: Caught in the Web*, directed by Annie Goldson (Gravitas Ventures, 2017).

PART TWO. BREAKING

1. "Google: End Piracy, Not Liberty," Google, archived at Wayback Machine, citing a capture dated January 18, 2012, https://web.archive.org/web/20120118084920/https://www.google.com/landing/takeaction/.

2. Josh Smith, "Spending Plus Online Clout Put Google in Lobbying Class of Its Own," *National Journal*, January 23, 2012, archived at Wayback Machine, citing a capture dated January 23, 2012, https://web.archive.org/web/20120123195234/http://www.nationaljournal.com/tech/spending-plus-online-clout-put-google-in-lobbying-class-of-its-own-20120123.

3. Edward Lee, *The Fight for the Future: How People Defeated Hollywood and Saved the Internet—for Now* (Chicago: Edward Lee, 2013). See also Bill D. Herman, *The Fight over Digital Rights: The Politics of Copyright and Technology* (Cambridge: Cambridge University Press, 2013).

4. Smith, "Spending Plus Online Clout."

5. Eric Schmidt, quoted in Gautham Nagesh, "Google Chairman Says Online Piracy Bill Would 'Criminalize' the Internet," *The Hill*, December 12, 2012, archived at Wayback Machine, citing a capture dated January 13, 2012, https://web.archive.org/web/20120113034439/http://thehill.com/blogs/hillicon-valley/technology/198777-google-chairman-says-online-piracy-bill-would-criminalize-linking.

6. Schmidt, quoted in Nagesh, "Google Chairman."

7. Sergey Brin, quoted in Jennifer Martinez and Michelle Quinn, "Will SOPA Kill the Internet?," *Politico*, December 17, 2011, https://www.politico.com/story/2011/12/will-sopa-kill-the-internet-070554.

8. "URGENT ACTION NEEDED: New Internet Blacklist Bill Could Shut Down Twitter and YouTube!," Demand Progress, archived at Wayback Machine, citing a capture dated October 28, 2011, https://web.archive.org/web/20111028204811/http://demandprogress.org/.

9. Steve Tepp, "Reality Checks: Don't Be Fooled by Anti-IP's Scare Tactics," Coalition against Counterfeiting and Piracy, October 28, 2011, archived at Wayback Machine, citing a capture dated January 4, 2012, https://web.archive.org/web/20120104133923/http://thecacp.com/blogs/reality-checks-don%E2%80%99t-be-fooled-anti-ip%E2%80%99s-scare-tactics.

10. Dong Ngo, "Italian-Language Wikipedia Hidden, May Shut Down," *CNET News*, October 4, 2011, archived at Wayback Machine, citing a capture dated October 7, 2011, https://web.archive.org/web/20111007180323mp_/http://news.cnet.com/8301-1023_3-20115615-93/italian-language-wikipedia-hidden-may-shut-down/; Rebecca J. Rosen, "Italian Wikipedia Shuts Down in Protest of Wiretap Act," *Atlantic*, October 5, 2011, archived at Wayback Machine, citing a capture dated February 4, 2012, https://web.archive.org/web/20120204081100/https://www.theatlantic.com/technology/archive/2011/10/italian-wikipedia-shuts-down-in-protest-of-wiretap-act/246180/.

11. Rosen, "Italian Wikipedia Shuts Down."

12. "User talk:Jimbo Wales," Wikipedia, archived at Wayback Machine, citing a capture dated December 14, 2011, https://web.archive.org/web/20111214084720/https://en.wikipedia.org/wiki/User_talk:Jimbo_Wales#Request_for_Comment:_SOPA_and_a_strike.

13. Mike Fahmie and Jeremy Vinar, "THIS IS A COPYRIGHTED IMAGE," Imgur, archived at Wayback Machine, citing a capture dated February 4, 2012, https://web.archive.org/web/20120204052010/http://i.imgur.com/TD4Kq.jpg.

14. J. S., "VETO the SOPA Bill and Any Other Future Bills That Threaten to Diminish the Free Flow of Information," We the People: Your Voice in Government, whitehouse.gov, December 18, 2011, archived at Wayback Machine, citing a capture dated March 6, 2013, https://web.archive.org/web/20130306021537/https://petitions.whitehouse.gov/petition/veto-sopa-bill-and-any-other-future-bills-threaten-diminish-free-flow-information/g3W1BscR.

15. Danny Weitzner and Karen Kornbluh, "Agreement Reached on Internet Policymaking Principles," Obama White House Blog Archives, July 1, 2011, https://obamawhitehouse.archives.gov/blog/2011/07/01/agreement-reached-internet-policymaking-principles.

16. Office of the President of the United States, "International Strategy for Cyberspace: Prosperity, Security, and Openness in a Networked World," May 2011, https://obamawhitehouse.archives.gov/sites/default/files/rss_viewer/international_strategy_for_cyberspace.pdf.

17. Weitzner and Kornbluh, "Agreement Reached."

18. Weitzner and Kornbluh, "Agreement Reached."

19. J. S., "VETO the SOPA bill."

20. Trevor Timm, "How PIPA and SOPA Violate White House Principles Supporting Free Speech and Innovation," Electronic Frontier Foundation, January 16, 2012, archived at Wayback Machine, citing a capture dated January 18, 2012, https://web.archive.org/web/20120118013718/https://www.eff.org/deeplinks/2012/01/how-pipa-and-sopa-violate-white-house-principles-supporting-free-speech.

21. "Obama Administration Responds to the We the People Petitions on SOPA and Online Piracy," January 14, 2012, https://obamawhitehouse.archives.gov/blog/2012/01/14/obama-administration-responds-we-people-petitions-sopa-and-online-piracy.

22. Timm, "How PIPA and SOPA Violate White House Principles."

23. Quoted in Jenna Wortham and Somini Sengupta, "Bills to Stop Web Piracy Invite a Protracted Battle," January 15, 2012, archived at Wayback Machine, citing a capture dated January 18, 2012, https://web.archive.org/web/20120118063023/http://www.nytimes.com/2012/01/16/technology/web-piracy-bills-invite-a-protracted-battle.html?_r=1.

24. Clay Shirky, quoted in Jenna Wortham, "Public Outcry over Antipiracy Bills Began as Grass-Roots Grumbling," *New York Times*, January 19, 2012.

25. Chris Piascik, "Save the LOL Cats," January 18, 2012, https://www.flickr.com/photos/chrispiascik/6718984211/.

26. Eric Engleman, "Collapse of Anti-Piracy Bills Leaves Hollywood Seeking a Truce," *Bloomberg News*, January 26, 2012, https://www.bloomberg.com/news/articles/2012-01-23/collapse-of-anti-piracy-bills-leaves-hollywood-seeking-a-truce.

CHAPTER FOUR. CRACKING THE CODE

1. Leonardo Chiariglione, "Press Release: Worldwide Recording Industry and Technology Companies Kick Off Work of Secure Digital Music Initiative," Secure Digital Music Initiative, February 26, 1999, archived at Wayback

Machine, citing a capture dated July 13, 2000, https://web.archive.org/web/20000713001746/http://www.sdmi.org/pr/LA_Feb_26_1999_PR.htm.

2. "Secure Digital Music Initiative (SDMI) Fact Sheet," Secure Digital Music Initiative, archived at Wayback Machine, citing a capture dated January 18, 2000, https://web.archive.org/web/20000118203024/http://www.sdmi.org/public_doc/FinalFactSheet.htm.

3. "Leonardo—One Time Coding Researcher, Standardisation Veteran and Entrepreneur," Leonardo Chiariglione, https://leonardo.chiariglione.org/; emphasis in original.

4. Tarleton Gillespie, *Wired Shut: Copyright and the Shape of Digital Culture* (Cambridge, MA: MIT Press, 2007).

5. Margaret Quan, "SDMI Releases Portable-Player Spec," *EE Times,* July 13, 1999, https://www.eetimes.com/sdmi-releases-portable-player-spec/.

6. Leonardo Chiariglione, quoted in Quan, "SDMI Releases Portable-Player Spec."

7. Leonardo Chiariglione, "An Open Letter to the Digital Community," Secure Digital Music Initiative, September 6, 2000, archived at Wayback Machine, citing a capture dated February 16, 2004, https://web.archive.org/web/20040216013811/http://www.sdmi.org/pr/OL_Sept_6_2000.htm.

8. "Declaration of Edward Felten in *Felten v. RIAA* (August 13, 2001)," reposted by Electronic Frontier Foundation, archived at Wayback Machine, citing a capture dated September 12, 2016, https://web.archive.org/web/20160912132838/https://w2.eff.org/IP/DMCA/Felten_v_RIAA/20010813_felten_decl.html.

9. "Declaration of Felten."

10. "Click-Through Agreement," Secure Digital Music Initiative, cited in "Declaration of Felten."

11. "Click-Through Agreement."

12. "Declaration of Felten."

13. "Click-Through Agreement."

14. "Declaration of Felten."

15. "Click-Through Agreement."

16. Universal City Studios, Inc. v. Reimerdes, 111 F.Supp. 2d 294 (S.D.N.Y. 2000). A useful overview of the case is found in Lawrence Lessig, *The Future of Ideas: The Fate of Commons in a Connected World* (New York: Random House, 2001), 187–190.

17. "Declaration of Felten."

18. Anonymous peer reviewer, quoted in "Declaration of Felten."

19. Email from Joseph Winograd to Edward Felten, quoted in "EFF Complaint: Felten v. RIAA (June 6, 2001)," reposted by Electronic Frontier Foundation, archived at Wayback Machine, citing a capture dated October 20, 2012, https://web.archive.org/web/20121020171237/https://w2.eff.org/IP/DMCA/Felten_v_RIAA/20010606_eff_felten_complaint.html.

20. Letter from Matthew Oppenheim to Edward Felten, quoted in "EFF Complaint."

21. Oppenheim to Felton; "EFF Complaint."

22. "Declaration of Felten."

23. "Declaration of Felten."

24. "EFF Complaint."

25. Edward Felten, "Statement at the Fourth International Information Hiding Workshop," April 26, 2001, archived at Wayback Machine, citing a capture dated October 20, 2012, https://web.archive.org/web/20121020171224/https://w2.eff.org/IP/DMCA/Felten_v_RIAA/20010426_felten_message.html.

26. Lessig, *Future of Ideas*, 190.

27. "Declaration of Eleanor Young (USENIX Association) in *Felten v. RIAA* (August 13, 2001)," reposted by Electronic Frontier Foundation, archived by Wayback Machine, citing a capture dated October 20, 2012, https://web.archive.org/web/20121020171451/https://w2.eff.org/IP/DMCA/Felten_v_RIAA/20010813_usenix_decl.html; Peter H. Salus, "USENIX History," USENIX, April 2005, https://www.usenix.org/legacy/about/history/index.html.

28. "Declaration of Young."

29. "Declaration of Young."

30. "Declaration of Young."

31. "Declaration of Young."

32. "EFF Complaint."

33. "EFF Complaint."

34. "EFF Complaint."

35. "EFF Complaint."

36. "First Amended Felten Complaint, June 26, 2001," reposted by Electronic Frontier Foundation, archived by Wayback Machine, citing a capture dated October 20, 2012, https://web.archive.org/web/20121020171220/https://w2.eff.org/IP/DMCA/Felten_v_RIAA/20010626_eff_felten_amended_complaint.html.

37. Memorandum in Support to Dismiss, at Felton v. RIAA, 1. https://web.archive.org/web/20240121151415/https://w2.eff.org/IP/DMCA/Felten_v_RIAA/20010712_riaa_mtd_memo.pdf.

38. Memorandum in Support to Dismiss, 1–2.

39. "Statement by RIAA's Cary Sherman on Felten Lawsuit," Electronic Frontier Foundation, June 6, 2001, archived by Wayback Machine, citing a capture dated October 20, 2012, https://web.archive.org/web/20121020171253/https://w2.eff.org/IP/DMCA/Felten_v_RIAA/20010606_riaa_statement.html.

40. "EFF Reply to RIAA Statement regarding Felten Case," Electronic Frontier Foundation, June 6, 2001, archived by Wayback Machine, citing a capture dated October 20, 2012, https://web.archive.org/web/20121020171528/https://w2.eff.org/IP/DMCA/Felten_v_RIAA/20010606_eff_felten_riaa_reply.html.

41. "Scientists Support Professor's Copyright Law Challenge: Electronic Frontier Foundation Exposes 'Chilling Effect,' " Press Release, Electronic Frontier Foundation, August 13, 2001, archived at Wayback Machine, citing a capture dated November 30, 2008, https://web.archive.org/web/20081130073859/http://w2.eff.org/IP/DMCA/Felten_v_RIAA/20010813_eff_felten_pr.html.

42. "Declaration of Felten."

43. Edward Felten, quoted in "Scientists Support Professor's Copyright Law Challenge."

44. Henry Cowles, *The Scientific Method: An Evolution of Thinking from Darwin to Dewey* (Cambridge, MA: Harvard University Press, 2020).

45. DEF CON 9 main page, DEF CON 9, March 27, 2001, archived at Wayback Machine, citing a capture dated April 1, 2001, https://web.archive.org/web/20010401061329/http://www.defcon.org/html/defcon-9-pre.html.

46. DEF CON 9 events page, DEF CON 9, January 27, 2001, archived at Wayback Machine, citing a capture dated April 14, 2001, https://web.archive.org/web/20010414022946/http://www.defcon.org/html/defcon-9-events.html.

47. "DEF CON 2001: Speakers," DEF CON 9, July 6, 2001, archived at Wayback Machine, citing a capture dated October 12, 2001, https://web.archive.org/web/20011012043813/http:/www.defcon.org/html/defcon-9-speakers.html#Dmitry%20Sklyarov.

48. "DEF CON 2001: Speakers."

49. Company profile, ElcomSoft, archived at Wayback Machine, citing a capture dated August 6, 2001, https://web.archive.org/web/20010806084854/http://www.elcomsoft.com/company.html; "ElcomSoft.com: Order Software," ElcomSoft, archived at Wayback Machine, citing a capture dated August 6, 2001, https://web.archive.org/web/20010806092349/http://www.elcomsoft.com/order.html.

50. Criminal Complaint and Affidavit in Support of Complaint, at United States v. Sklyarov (July 7, 2001), 5, reposted by Electronic Frontier Foundation, archived at Wayback Machine, citing a capture dated November 15, 2001, https://web.archive.org/web/20011115080301/http://www.eff.org/IP/DMCA/US_v_Sklyarov/20010707_complaint.pdf.

51. Criminal Complaint at U.S. v. Sklyarov, reposted by Electronic Frontier Foundation, archived at Wayback Machine, citing a capture dated December 5, 2003, https://web.archive.org/web/20031205005615/http://www.eff.org/IP/DMCA/US_v_Elcomsoft/20010707_complaint.pdf.

52. Criminal Complaint.

53. ElcomSoft.com, in Criminal Complaint, 5–6.

54. ElcomSoft.com, in Criminal Complaint, 6.

55. ElcomSoft.com, in Criminal Complaint, 6–7.

56. ElcomSoft.com, in Criminal Complaint.

57. Elcomsoft.com, in Criminal Complaint, 3.

58. Elcomsoft.com, in Criminal Complaint.

59. "US v. ElcomSoft & Sklyarov FAQ: What Is the History and Current Status of This Case?," Electronic Frontier Foundation, February 19, 2002, https://www.eff.org/pages/us-v-elcomsoft-sklyarov-faq#Status.

60. "DoJ Press Release on Sklyarov Prosecution," Electronic Frontier Foundation, July 17, 2001, archived at Wayback Machine, citing a capture dated November 8, 2001, https://web.archive.org/web/20011108181128/http://www.eff.org/IP/DMCA/US_v_Sklyarov/20010717_doj_sklyarov_pr.html.

61. "DoJ Press Release," Department of Justice, July 17, 2001, capture dated July 31, 2023, https://web.archive.org/web/20230731043311/https://w2.eff.org/IP/DMCA/US_v_Elcomsoft/20010717_doj_sklyarov_pr.html.

62. "FBI Arrests Programmer in Las Vegas," Electronic Frontier Foundation, July 17, 2001, archived at Wayback Machine, citing a capture dated November 8, 2001, https://web.archive.org/web/20011108181253/http://www.eff.org/IP/DMCA/US_v_Sklyarov/20010717_eff_sklyarov_pr.html.

63. Shari Steele, "Statement on the Arrest of Dmitry Sklyarov," Electronic Frontier Foundation, July 18, 2001, archived at Wayback Machine, citing a capture dated November 8, 2001, https://web.archive.org/web/20011108181115/http://www.eff.org/IP/DMCA/US_v_Sklyarov/20010718_eff_sklyarov_statement.html.

64. "FBI Arrests Programmer in Las Vegas."

65. "Attend One of the Monday, July 23, 2001 Demonstration Rallies!," Boycott Adobe, archived at Wayback Machine, citing a capture dated August 3, 2001, https://web.archive.org/web/20010803131327/http://www.boycottadobe.com/pages/rallies.html.

66. "Attend One of the Monday, July 23, 2001 Demonstration Rallies!"

67. MarginalHacks, "dmca: MVC-152S.JPG," image of Adobe protest for Dmitry Sklyarov, July 31, 2001, archived at Wayback Machine, citing a capture dated January 5, 2002, https://web.archive.org/web/20020105163830/http://64.81.65.40/dmca/tn/MVC-152S.JPG.html; MarginalHacks, "dmca-MVC-112S.JPG," image of Adobe protest for Dmitry Sklyarov, July 31, 2001, archived at Wayback Machine, citing a capture dated October 12, 2001, https://web.archive.org/web/20011012104503/http://64.81.65.40/dmca/tn/MVC-112S.JPG.html; MarginalHacks, "dmca-MVC-118S.JPG," image of Adobe protest for Dmitry Sklyarov, July 31, 2001, archived at Wayback Machine, citing a capture dated December 5, 2001, accessed https://web.archive.org/web/20011205191206/http://64.81.65.40/dmca/tn/MVC-118S.JPG.html; MarginalHacks, "dmca MVC-118S.JPG"; MarginalHacks, "dmca-MVC-151S.JPG," image of Adobe protest for Dmitry Sklyarov, July 31, 2001, archived at Wayback Machine, citing a capture dated January 5, 2002, https://web.archive.org/web/20020105161700/http://64.81.65.40/dmca/tn/MVC-151S.JPG.html.

68. MarginalHacks, "dmca-MVC-157S.JPG," image of Adobe protest for Dmitry Sklyarov, July 31, 2001, archived at Wayback Machine, citing a capture dated January 5, 2002, https://web.archive.org/web/20020105231353/http://64.81.65.40/dmca/tn/MVC-157S.JPG.html.

69. "20010723-set2_41_2048x1536.jpg," image of Adobe protest for Dmitry Sklyarov, archived at Wayback Machine, citing a capture dated May 15, 2006, https://web.archive.org/web/20060515063751/http://primates.ximian.com/~peter/free-sklyarov/images/20010723-set2_41_2048x1536.jpg.

70. "Free Dmitry Resources for Boston," Free Sklyarov, archived at Wayback Machine, citing a capture dated August 17, 2003, https://web.archive.org/web/20030817051413/http://lesser-magoo.lcs.mit.edu/~cananian/sklyarov/.

71. "Free Dmitry Resources for Boston."

72. "Free Dmitry Resources for Boston."

73. Photos are available at Vladimir Smolyakov, "[First Rally] photos," Porsche Russia, archived at Wayback Machine, citing a capture dated April 9, 2002, https://web.archive.org/web/20020409045416/http://www.porsche.ru/dmitry/1st_rally_photos.html.

74. Ilya V. Vasilyev, "Moscow Demonstration Success," Porsche Russia, archived at Wayback Machine, citing a capture dated April 15, 2002, https://web.archive.org/web/20020415052236/http://www.porsche.ru/dmitry/1st_rally_report.html.

75. Ilya V. Vasilyev, "Moscow Protest: SUCCESS!," message to free-sklyarov newsgroup, August 30, 2001, archived at Wayback Machine, citing a capture dated April 29, 2003, https://web.archive.org/web/20030429000812/http://zork.net/pipermail/free-sklyarov/2001-August/003992.html.

76. Ilya V. Vasilyev, "E-Protest," Porsche Russia, August 30, 2001, archived at Wayback Machine, citing a capture dated April 15, 2002, https://web.archive.org/web/20020415044947/http://www.porsche.ru/dmitry/eprotest_report.html.

77. Vasilyev, "Moscow Protest."

78. "US v. ElcomSoft & Sklyarov FAQ."

79. Matt Berger, "Jury: ElcomSoft Not Guilty," *Network World*, December 17, 2002, https://www.networkworld.com/article/2339223/jury—elcomsoft-not-guilty.html.

CHAPTER FIVE. BROKEN VOTING MACHINES

1. "The Gravediggers: How Eliminalia, a Spanish Reputation Management Firm, Buries the Truth," Forbidden Stories, February 17, 2023, https://forbiddenstories.org/the-gravediggers-eliminalia/.

2. Meg Leta Jones, *Ctrl+Z: The Right to Be Forgotten* (New York: New York University Press, 2016).

3. Shawn Boburg, "Eliminalia Created Fake News, Bogus Legal Complaints: Leaked Files Reveal Reputation-Management Firm's Deceptive Tactics," *Washington Post*, February 17, 2023.

4. "Free Culture Swarthmore: History," Swarthmore College Computing Society, https://www.sccs.swarthmore.edu/org/scdc/history.html.

5. Sam Williams, *Free as in Freedom: Richard Stallman's Crusade for Free Software* (Sebastopol, CA: O'Reilly Media, 2002).

6. Williams, *Free as in Freedom.*

7. Christopher M. Kelty, *Two Bits: The Cultural Significance of Free Software* (Durham, NC: Duke University Press, 2008).

8. "Some Rights Reserved," Creative Commons, capture dated October 20, 2004, https://web.archive.org/web/20041020235420/https://creativecommons.org/about/history.

9. "About SCCS," Swarthmore College Computing Society, archived at Wayback Machine, citing a capture dated March 13, 2002, https://web.archive.org/web/20020313021502/http://www.sccs.swarthmore.edu/aboutsccs.html.

10. "Free Culture Swarthmore: History."

11. "Free Culture Swarthmore: History."

12. Lawrence Lessig, *Free Culture: How Big Media Uses Technology and the Law to Lock Down Culture and Control Creativity* (New York: Penguin, 2004).

13. Bev Harris, "Inside a U.S. Election Vote Counting Program," *Scoop,* July 8, 2003, archived at Wayback Machine, citing a capture dated September 15, 2004, https://web.archive.org/web/20040915083228/http://www.scoop.co.nz/mason/stories/HL0307/S00065.htm.

14. Harris, "Inside a U.S. Election Vote Counting Program." See also Bev Harris, *Black Box Voting: Ballot Tampering in the 21st Century* (N.p.: Talion, 2004).

15. Bush v. Gore, 531 U.S. 98 (2000).

16. Help America Vote Act of 2002, Pub. L. No. 107-252, 116 Stat. 1666 (2002).

17. Harris, "Inside a U.S. Election Vote Counting Program."

18. Message from "John Dee," reproduced as Exhibit D in Declaration of Cindy A. Cohn in Support of Plaintiffs' Motion for Summary Judgment, in Online Policy Group v. Diebold, Inc., No. C-03-04913 JF, February 9, 2004, 11, reposted by Electronic Frontier Foundation, archived at Wayback Machine, citing a capture dated June 23, 2011, https://web.archive.org/web/20110623215639/http://www.eff.org/files/filenode/OPG_v_Diebold/Decl_of_CAC_w_Exhs.pdf.

19. Ian S. Piper, "RE: Memory Card Checksum Errors (Was: 2000 November Election)," email to Global Elections System Support, January 18, 2001, archived at Wayback Machine, citing a capture dated May 21, 2006, https://web.archive.org/web/20060521230050/http://chroot.net/s/lists/support.w3archive/200101/msg00068.html.

20. Piper, "RE: Memory Card Checksum Errors."

21. Message from "John Dee," Exhibit E in Declaration of Cohn, 13.

22. Message from "John Dee," Exhibit F in Declaration of Cohn, 15.

23. A handful of these emails are available today at "Index of Diebold Mailing Lists," archived at Wayback Machine, citing a capture dated November 19, 2005, https://web.archive.org/web/20051119223117/http://chroot.net/s/lists/; Plaintiff's Notice of Motion and Motion for Summary Judgment, February 9, 2004, OPG v. Diebold; Keith Ervin, "Elections Chief Tightens Vote Security," *Seattle Times,* September 25, 2003, reposted by User FauxPrez to seattle.politics Narkive newsgroup thread "More on Diebold 'Frauds 'R' Us' Election Systems," https://seattle.politics.narkive.com/tBIrGiUu/more-on-diebold-frauds-r-us-election-systems.

24. One version of SCDC's post is available at "Diebold Memos Reposted on Swarthmore Servers!," Swarthmore Coalition for the Digital Commons, November 25, 2003, archived at Wayback Machine, citing a capture dated December 3, 2003, https://web.archive.org/web/20031203115331/http://www.sccs.swarthmore.edu/org/scdc/.

25. Adrian John, *Piracy: The Intellectual Property Wars from Gutenberg to Gates* (Chicago: University of Chicago Press, 2011).

26. Ralph Jocke, "RE: Copyright Infringement," letter to Judy Downing, October 9, 2003, in Declaration of Nelson Chu Pavlosky in Support of Plaintiff's Application for Temporary Restraining Order and Preliminary Injunction, OPG v. Diebold, reposted by Electronic Frontier Foundation, archived at Wayback Machine, citing a capture dated March 5, 2011, https://web.archive.org/

web/20110305071728/http://www.eff.org/files/filenode/OPG_v_Diebold/Pavlosky.pdf.

27. Why War? staff, "Excerpts from the Diebold Documents," in "Targeting Diebold with Electronic Civil Disobedience," Why War?, archived at Wayback Machine, citing a capture dated December 4, 2003, https://web.archive.org/web/20031204224431/http:/www.why-war.com/features/2003/10/diebold.html#excerpts.

28. Supplemental Declaration of Luke Thomas Smith in Support [of] Application for Primary Injunction, November 6, 2003, OPG v. Diebold, reposted by Electronic Frontier Foundation, archived at Wayback Machine, citing a capture dated March 5, 2011, https://web.archive.org/web/20110305071617/http://www.eff.org/files/filenode/OPG_v_Diebold/20031107_supp_dec_smith.pdf.

29. Why War? staff, "Excerpts."

30. Why War? staff, "Excerpts."

31. Declaration of Luke Thomas Smith in Support of Plaintiffs' Application for Temporary Restraining Order and for Preliminary Injunction, November 2, 2003, OPG v. Diebold, reposted by Electronic Frontier Foundation, archived at Wayback Machine, citing a capture dated March 5, 2011, https://web.archive.org/web/20110305071713/http://www.eff.org/files/filenode/OPG_v_Diebold/Smith.pdf.

32. Why War? staff, "Excerpts."

33. Luke Thomas Smith and Nelson Chu Pavlosky, letter to Judy Downing, November 6, 2003, Exhibit A in Supplemental Declaration of Luke Thomas Smith in Support Application for Preliminary Injunction, OPG v. Diebold, reposted by Electronic Frontier Foundation, archived at Wayback Machine, citing a capture dated March 5, 2011, https://web.archive.org/web/20110305071617/http://www.eff.org/files/filenode/OPG_v_Diebold/20031107_supp_dec_smith.pdf.

34. Paul Hales, " 'Civil Disobedience' Campaign Targets Diebold," Inquirer, October 22, 2003, archived at Wayback Machine, citing a capture dated December 2, 2003, https://web.archive.org/web/20031202173608/http://www.theinquirer.net/?article=12261; Kim Zetter, "Students Fight E-Vote Firm," Wired News, October 21, 2003, https://www.wired.com/2003/10/students-fight-e-vote-firm/; Don Russel, "Swarthmore Groups Told to Nix Links: Firm Says Memos Were Stolen by Hacker," Philadelphia Daily News, October 23, 2003, archived at Wayback Machine, citing a capture dated February 19, 2004, https://web.archive.org/web/20040219142532/http://www.philly.com/mld/dailynews/news/local/7080760.htm; Rachel Konrad, "Diebold Threatens Publisher of Leaked Documents," Associated Press, October 27, 2003, reposted by Why War?, archived at Wayback Machine, citing a capture dated November 27, 2003, https://web.archive.org/web/20031127043802/http://www.why-war.com/news/2003/10/27/dieboldt.html.

35. Andrea Foster, "Swarthmore Shuts Down Web Sites of Students Publicizing Company's Voting-Machine Memos," Chronicle of Higher Education, October 27, 2003, reposted to Why War?, archived at Wayback Machine, citing a capture

dated May 5, 2004, https://web.archive.org/web/20040505111540/http://www.why-war.com/news/2003/10/27/swarthmo.html.

36. Bev Harris, Homepage, Black Box Voting: Ballot-Tampering in the 21st Century, archived at Wayback Machine, citing a capture dated November 20, 2003, https://web.archive.org/web/20031120181931/http://www.blackboxvoting.org/.

37. "Link to the Full Stash of Diebold Memos . . .," post to Indybay news section, September 29, 2003, archived at Wayback Machine, citing a capture dated December 2, 2003, https://web.archive.org/web/20031202192510/http://www.indybay.org/news/2003/09/1649419_comment.php.

38. Luke Smith, quoted in Zetter, "Students Fight E-Vote Firm."

39. Rebecca Mercuri, quoted in John Schwartz, "File Sharing Pits Copyright against Free Speech," *New York Times*, November 3, 2003.

40. Marc Carrel, quoted in Kim Zetter, "Calif. Halts E-Vote Certification," *Wired News*, November 3, 2003, https://www.wired.com/2003/11/calif-halts-e-vote-certification/.

41. Zetter, "Calif. Halts E-Vote Certification."

42. "Staff: Wendy Seltzer," Electronic Frontier Foundation, archived at Wayback Machine, citing a capture dated March 6, 2005, https://web.archive.org/web/20050306154353/http://eff.org/about/staff/?f=wendy_seltzer.html; Homepage, Chilling Effects Clearinghouse, archived at Wayback Machine, citing a capture dated March 6, 2005, https://web.archive.org/web/20050306051416/http://www.chillingeffects.org/.

43. Wendy Seltzer, "Re: Diebold's Copyright Infringement Claim," message to Ralph E. Jocke sent via email and US mail, October 22, 2003, at Declaration of Wendy Seltzer in Support of Plaintiffs' Application for Temporary Restraining Order and for Preliminary Injunction, 4, OPG v. Diebold, reposted by Electronic Frontier Foundation, archived at Wayback Machine, citing a capture dated May 14, 2011, https://web.archive.org/web/20110514010818/http://www.eff.org/files/filenode/OPG_v_Diebold/Seltzer.pdf.

44. Seltzer, "Re: Diebold's Copyright Infringement Claim."

45. "Staff Bio: Cindy Cohn, Legal Director," Electronic Frontier Foundation, archived at Wayback Machine, citing a capture dated February 13, 2005, https://web.archive.org/web/20050213095243/http://www.eff.org/about/staff/?f=cindy_cohn.html.

46. "People," Stanford Center for Internet and Society, archived at Wayback Machine, citing a capture dated November 10, 2001, https://web.archive.org/web/20011110141846/http://cyberlaw.stanford.edu/people/index.shtml.

47. Plaintiffs' Memorandum of Points and Authorities in Support of Ex Parte Application for a Temporary Restraining Order and Order to Show Cause Re: Preliminary Injunction, 11, OPG v. Diebold, reposted by Electronic Frontier Foundation, archived at Wayback Machine, citing a capture dated May 14, 2011, https://web.archive.org/web/20110514011003/http://www.eff.org/files/filenode/OPG_v_Diebold/tro_brief.pdf.

48. Cindy Cohn, "Re: Online Policy group v. Diebold, No. C-03-4913 JF," letter to Jeremy Fogel, November 17, 2003, reposted by Electronic Frontier Foundation,

archived at Wayback Machine, citing a capture dated May 14, 2011, https://web.
archive.org/web/20110514011018/http://www.eff.org/files/filenode/OPG_v_
Diebold/031117_Cohn_letter.pdf.

49. Ralph E. Jocke, "Re: Copyright Infringement: Second Notice," letter to Benny
Ng, November 17, 2003, in Exhibit A of Supplemental Declaration of Benny Ng,
November 17, 2003, OPG v. Diebold, reposted by Electronic Frontier
Foundation, archived at Wayback Machine, citing a capture dated May 14, 2011,
https://web.archive.org/web/20110514011055/http://www.eff.org/files/filenode/
OPG_v_Diebold/Ng_suppl_decl.pdf.

50. Harper & Row Publishers, Inc., v. Nation Enters., 471 U.S. 539 (1985).

51. Opposition to Application for Temporary Restraining Order, November 4, 2003,
OPG v. Diebold, reposted by Electronic Frontier Foundation, archived at
Wayback Machine, citing a capture dated May 14, 2011, https://web.archive.org/
web/20110514010601/http://www.eff.org/files/filenode/OPG_v_Diebold/20031104-
Diebold-opp.pdf.

52. Tadayoshi Kohno, Adam Stubblefield, Aviel Rubin, and Dan Wallach, "Analysis of
an Electronic Voting System" (paper presented at IEEE Symposium on Security
and Privacy 2004, Oakland, CA, May 9–12, 2004), https://avirubin.com/vote.pdf.

53. "Kucinich Requests House Judiciary Committee Hearing on Diebold's Abuses of
Digital Millennium Copyright Act: Sends Letter to Chairman Sensenbrenner
and Ranking Member Conyers," Press Release, US House of Representatives:
Ohio 10th District, November 21, 2003, archived at Wayback Machine, citing a
capture dated December 24, 2003, https://web.archive.org/web/2003122
4212840/http://www.house.gov/apps/list/press/ohio_kucinich/031121jud
cmtediebold.html. See also "Voting Rights," US House of Representatives:
Congressman Dennis J. Kucinich of the 10th District of Ohio, archived at
Wayback Machine, citing a capture dated December 2, 2003, https://web.
archive.org/web/20031202220752/http://www.house.gov/kucinich/issues/
voting.htm.

54. "Diebold Backs Down, Won't Sue on Publication of Electronic Voting Machine
Flaws: Court Schedules Mediation and Hearing in Electronic Voting Case,"
Electronic Frontier Foundation, December 1, 2003, archived at Wayback
Machine, citing a capture dated December 4, 2003, https://web.archive.org/
web/20031204211724/http://www.eff.org/Legal/ISP_liability/OPG_v_
Diebold/20031201_eff_pr.php.

55. Robert J. Urosevich, "EFF: Diebold Withdrawal Letter to OPG from Robert
Urosevich, President," letter to Mr. Doherty of Diebold Election Systems,
December 3, 2003, in Exhibit D of Declaration of Adam Sand in Support of
Diebold's Motion for Summary Judgment, 18–19, OPG v. Diebold, January 12,
2004, archived at Wayback Machine, citing a capture dated May 14, 2011, https://
web.archive.org/web/20110514010542/http://www.eff.org/files/filenode/
OPG_v_Diebold/Diebold_Decl_of_Sand.pdf.

56. "Get the Diebold Memos Here!," Swarthmore Coalition for the Digital
Commons, December 9, 2003, in Declaration of Sand, 20–21.

57. Why War? staff, "Targeting Diebold with Electronic Civil Disobedience."

58. "Diebold Backs Down."

59. Wendy Seltzer, quoted in Mary Bridges, "Berkman Briefing: Diebold vs. The Bloggers," October 3, 2004, Berkman Klein Center, https://cyber.harvard.edu/ publications/2004/Diebold_vs_The_Bloggers.

60. "Alan Korn: Biography," Law Office of Alan Korn, http://www.alankorn.com/bio. html.

61. Memorandum in Support of Defendants' Motion for Summary Judgment, OPG v. Diebold, January 12, 2004, reposted by Electronic Frontier Foundation, archived at Wayback Machine, citing a capture dated May 14, 2011, https://web. archive.org/web/20110514011059/http://www.eff.org/files/filenode/OPG_v_ Diebold/Diebold_MSJ_Memo.pdf.

62. Online Policy Group v. Diebold, Inc., 337 F.Supp.2d 1195, 1204–1205 (N.D. Cal. 2004).

63. Chris Nolan, "Diebold Debacle Signals Need for a Paper Trail," *eWeek*, October 6, 2004, https://www.eweek.com/news/diebold-debacle-signals-need-for-a- paper-trail/.

64. "Good Intentions, Bad Technology: High-Tech Voting Machines Are Making Things Worse, Not Better," *Economist*, January 22, 2004, https://www.economist. com/united-states/2004/01/22/good-intentions-bad-technology.

65. "Good Intentions, Bad Technology."

66. Dan Verton, "E-Vote at Risk: Despite Vendor Assurances, Researchers Remain Concerned about the Security and Reliability of Electronic Voting Systems," *Computerworld*, October 18, 2004, 25, archived by Google Books, https://books. google.com/books?id=l2jybsH3VfwC&lpg=PP1&pg=PA25.

67. "ODNI Announces Senior Leadership Positions," Press Release, Office of the Director of National Intelligence, October 31, 2005, archived at Wayback Machine, citing a capture dated September 23, 2006, https://web.archive.org/ web/20060923133757/http://www.dni.gov/press_releases/20051031_release. htm.

68. Michael A. Wertheimer, "Trusted Agent Report Diebold AccuVote-TS Voting System," January 20, 2004, 3, http://people.csail.mit.edu/rivest/voting/ reports/2004-01-20%20RABA%20evaluation%20of%20Diebold%20AccuVote. pdf.

69. Wertheimer, "Trusted Agent Report."

70. John Schwartz, "Security Poor in Electronic Voting Machines, Study Warns," *New York Times*, January 29, 2004.

71. *Hacking Democracy*, directed by Simon Ardizzone and Russel Michaels (HBO, 2006).

72. Barney Warf, "Voting Technologies and Residual Ballots in the 2000 and 2004 Presidential Elections," *Political Geography* 25, no. 5 (2006): 530–556.

73. Christina A. Cassidy, "EXPLAINER: Voting Systems Reliable, Despite Conspiracies," Associated Press, October 4, 2022, https://apnews.com/ article/2022-midterm-elections-technology-voting-donald-trump-campaigns- 46c9cf208687636b8eaa1864c35ab300; Tami Abdollah, "US Designates Election Infrastructure as 'Critical,' " Associated Press, January 6, 2017, https://

apnews.com/united-states-government-64a7228c974d43009cdfc2b9876632ob; "Russian Interference in 2016 U.S. Elections," Most Wanted, Federal Bureau of Investigation, https://www.fbi.gov/wanted/cyber/russian-interference-in-2016-u-s-elections.

CHAPTER SIX. SPONGEBOB GOES TO COURT

1. Fred von Lohmann, "Unfairly Caught in Viacom's Dragnet? Let Us Know!," Electronic Frontier Foundation, February 8, 2007, archived at Wayback Machine, citing a capture dated February 19, 2007, https://web.archive.org/web/20070219121838/http://www.eff.org/deeplinks/archives/005109.php.

2. James F. Moore (@venturecapitalview), "Sunday Nite Dinner at Redbones in Somerville, Mass," YouTube video of dinner at Redbones Bar, September 24, 2006, archived at Wayback Machine, citing a capture dated February 25, 2007, https://web.archive.org/web/20070225135348/http://www.youtube.com/watch%3Fv%3DQUzOP42dgiI.

3. Moore, "Sunday Nite Dinner." "Viacom" lacks a "V" in the original.

4. "RIAA v. The People: Two Years Later," Electronic Frontier Foundation, archived at Wayback Machine, citing a capture dated February 23, 2007, https://web.archive.org/web/20070223063954/http://www.eff.org/IP/P2P/RIAAatTWO_FINAL.pdf.

5. Lawrence Lessig, *Free Culture: How Big Media Uses Technology and the Law to Lock Down Culture and Control Creativity* (New York: Penguin, 2004).

6. Dennis Roddy, in "RIAA v. The People," 4.

7. Cory Doctorow, "Viacom Terrorizes YouTube with Bullshit DMCA Notices," *Boingboing*, February 3, 2007, archived at Wayback Machine, citing a capture dated February 20, 2007, https://web.archive.org/web/20070220185457/http://www.boingboing.net/2007/02/03/viacom_terrorizes_yo.html.

8. Complaint for Declaratory and Injunctive Relief and Damages, Viacom Int'l v. YouTube, Inc., March 12, 2007, 13, archived at Wayback Machine, citing a capture dated July 10, 2007, https://web.archive.org/web/20070710024614/https://www.eff.org/legal/cases/viacom_v_google/ViacomYouTubeComplaint3-12-07.pdf.

9. Complaint for Declaratory and Injunctive Relief and Damages, 13.

10. Opinion and Order, Viacom Int'l, Inc., v. YouTube, Inc., and The Football Ass'n Premier League, Ltd., v. YouTube, Inc., Doc. 110 (S.D.N.Y. July 1, 2008), 22n12, https://cases.justia.com/federal/district-courts/new-york/nysdce/1:2007cv02103/302164/117/0.pdf.

11. Bloomberg Originals (@bloomberg), "Google Says Viacom Secretly Uploaded Clips to YouTube: Video," YouTube video clip of Bloomberg, March 19, 2012, broadcast, March 23, 2012, https://youtu.be/3As61RajDyI.

12. Steve Chen, "The State of Our Video ID Tools," Google blog, June 14, 2007, https://googleblog.blogspot.com/2007/06/state-of-our-video-id-tools.html.

13. Kevin J. Delaney, "YouTube to Test Software to Ease Licensing Fights," *Wall Street Journal*, June 12, 2007, archived at Wayback Machine, citing a capture

dated March 12, 2015, https://web.archive.org/web/20150312061146/https://www.wsj.com/articles/SB11816129562693214.

14. David King, "Content ID Turns Three," YouTube blog, December 2, 2010, https://blog.youtube/news-and-events/content-id-turns-three/.

15. Emma Baccellieri, "Mud Maker: The Man behind MLB's Essential Secret Sauce," *Sports Illustrated,* August 7, 2019, https://www.si.com/mlb/2019/08/07/baseball-mud-rawlings.

16. Jordyn Holman, "Silicon Valley Is Using Trade Secrets to Hide Its Race Problem," Bloomberg, February 13, 2019, https://www.bloomberg.com/news/articles/2019-02-13/silicon-valley-is-using-trade-secrets-to-hide-its-race-problem.

17. Reply, cited in Opinion and Order, Viacom v. YouTube, 10.

18. Amended Stipulated Pre-Trial Protective Order, Viacom v. YouTube, January 24, 2008, 4, https://cases.justia.com/federal/district-courts/new-york/nysdce/1:2007cv02103/302164/76/0.pdf.

19. Amended Stipulated Pre-Trial Protective Order, 5.

20. Amended Stipulated Pre-Trial Protective Order, 8–9.

21. Viacom Int'l, Inc. v. YouTube, Inc., 718 F.Supp. 2d 514 (S.D.N.Y. 2010).

22. Shane Greenstein, *How the Internet Became Commercial: Innovation, Privatization, and the Birth of a New Network* (Princeton, NJ: Princeton University Press, 2015).

23. Shoshana Zuboff, *The Age of Surveillance Capitalism: The Fight for a Human Future at the New Frontier of Power* (New York: Public Affairs, 2019).

24. Amended Stipulated Pre-Trial Protective Order, 12.

25. Amended Stipulated Pre-Trial Protective Order, 13.

26. 18 U.S.C. § 2710, cited in Kurt Opsahl, "Court Ruling Will Expose Viewing Habits of YouTube Users," Electronic Frontier Foundation, July 2, 2008, https://www.eff.org/deeplinks/2008/07/court-ruling-will-expose-viewing-habits-youtube-us.

27. Opsahl, "Court Ruling."

28. Viacom v. YouTube.

29. Viacom v. YouTube.

30. Opening Brief for Plaintiffs-Appellants, Viacom v. YouTube, Doc. 59, December 3, 2010, 39, https://cases.justia.com/federal/appellate-courts/ca2/10-3270/59/0.pdf.

31. Viacom Int'l, Inc., v. YouTube, Inc., 1:07-cv-02103, Doc. 452 (S.D.N.Y. April 18, 2013), https://www.docketalarm.com/cases/New_York_Southern_District_Court/1-07-cv-02103/Viacom_International_Inc._et_al_v._Youtube_Inc._et_al/452/.

PART THREE. TRANSFORMING

1. Safiya U. Noble, *Algorithms of Oppression: How Search Engines Reinforce Racism* (New York: New York University Press, 2018).

2. Grace Mayer, "OpenAI Cofounder Responds to Elon Musk's Criticism That ChatGPT Is Too 'Woke': 'We Made a Mistake,' " *Business Insider,* March 9, 2023, https://www.businessinsider.com/

openai-cofounder-greg-brockman-responds-elon-musk-woke-criticism-chatgpt-2023-3.

3. OpenAI, "Usage Policies," capture dated February 2, 2024, https://web.archive.org/web/20240202012538/https://openai.com/policies/usage-policies.

4. "Wikimania Gets Social," Barbican, August 11, 2014, archived at Wayback Machine, citing a capture dated March 4, 2016, https://web.archive.org/web/20160304071826/http://blog.barbican.org.uk/2014/08/wikimania-gets-social/.

5. Naruto v. Slater, No. 16-15469 (9th Cir. 2018), at 2.

6. Naruto v. Slater, 17.

7. Roe v. Wade, 410 U.S. 113 (1973); Davis v. Michigan Dep't of Treasury, 489 U.S. 803, 809 (1989).

8. All quotations from Burrow-Giles Lithographic Co. v. Sarony, 111 U.S. 53, 58–59 (1884).

9. US Copyright Office, *Compendium of U.S. Copyright Office Practices*, 3rd ed. (Washington, DC, 2021), § 313.2.

10. Robert Kasunic, "Re: Zarya of the Dawn (Registration # VAu001480196)," letter to Van Lindberg, February 21, 2023, https://fingfx.thomsonreuters.com/gfx/legaldocs/klpygnkyrpg/AI%20COPYRIGHT%20decision.pdf.

11. US Copyright Office Review Board, "Re: Second Request for Reconsideration for Refusal to Register a Recent Entrance to Paradise," letter to Ryan Abbott, February 14, 2022, https://www.copyright.gov/rulings-filings/review-board/docs/a-recent-entrance-to-paradise.pdf.

12. Andres Gaudamuz, "Artificial Intelligence and Copyright," *WIPO Magazine*, October 2017, https://www.wipo.int/wipo_magazine/en/2017/05/article_0003.html.

13. Copyright, Designs, and Patent Act (UK), cited in Gaudamuz, "Artificial Intelligence and Copyright."

14. Holly Guenther, "About," Kimchi Kawaii, https://www.kimchikawaii.com/about.

15. Damir Yalalov, "Midjourney and Dall-E Artist Styles Dump with Examples: 130 Famous AI Painting Techniques," Metaverse Post, August 11, 2022, updated July 18, 2024, https://mpost.io/midjourney-and-dall-e-artist-styles-dump-with-examples-130-famous-ai-painting-techniques/.

16. Yalalov, "Midjourney and Dall-E Artist Styles Dump."

17. Kyle Chayka, "Is A.I. Art Stealing from Artists?," *New Yorker*, February 10, 2023, https://www.newyorker.com/culture/infinite-scroll/is-ai-art-stealing-from-artists.

CHAPTER SEVEN. RAPPERS TO THE RESCUE

1. Skyywalker Records, Inc. v. Navarro, 739 F. Supp. 578 (S.D. Fla. 1990).

2. US Copyright Office, "Mechanical License Royalty Rates," March 2024, https://copyright.gov/licensing/m200a.pdf.

3. Acuff-Rose Music, Inc. v. Campbell, 754 F. Supp. 1150 (M.D. Tenn. 1991).

4. The 2 Live Crew, *As Clean as They Wanna Be*, compact disc, June 15, 1989, Luke Skyywalker Records CDXR108.

5. "Memorandum," Acuff-Rose v. Campbell, No. 3:90-0524, 754 F. Supp. 1150 (M.D. Tenn. 1991), https://law.justia.com/cases/federal/district-courts/FSupp/754/1150/2353338/.

6. 17 U.S.C. § 107, "Limitations on Exclusive Rights: Fair Use."

7. Harper & Row Publishers, Inc., v. Nation Enters., 471 U.S. 539 (1985); Sony Corp. of America v. Universal City Studios, Inc., 464 U.S. 417 (1984).

8. Acuff-Rose Music, Inc. v. Campbell, 972 F.2d 1429, 1437 (6th Cir. 1992).

9. Campbell v. Acuff-Rose Music, Inc., 510 U.S. 569, 579 (1994).

10. Campbell v. Acuff-Rose, 579.

11. Complaint, Kelly v. Arriba Soft Corp., 99-CV-00560-GLT-AN, US District Court for the Central District of California, 2002, 6.

12. Complaint, 7.

13. "Arriba Soft Introduces First Software Tool Enabling PC Users to Experience and Harness Rich Media," Press Release, Arriba Soft, July 27, 1998, archived at Wayback Machine, citing a capture dated October 13, 1999, https://web.archive.org/web/19991013234617/http://www.arribasoft.com/products/express/press/aerel_2.htm.

14. "Arriba Soft Introduces First Software Tool." See also "Media Asset Management: A White Paper from Arriba Soft Corporation," Arriba Soft, archived at Wayback Machine, citing a capture dated February 20, 1999, https://web.archive.org/web/19990220151642/http://www.arribasoft.com/lounge/library/wp_mediaman.html.

15. "Arriba Express Overview: Arriba Express Media Management for Windows 95/98 & NT 4.0," Arriba Soft, archived at Wayback Machine, citing a capture dated April 27, 1999, https://web.archive.org/web/19990427100118/http://www.arribasoft.com/products/express/overview.html.

16. Brief of Defendant-Appellee, 9th Circuit, 6–8; Brief for Plaintiff-Appellant, 10–12.

17. "Arriba in the News," Arriba Soft, archived at Wayback Machine, citing a capture dated April 27, 1999, https://web.archive.org/web/19990427094137/http://www.arribasoft.com/products/express/news.asp.

18. "Copyright/Terms of Use," Arriba Vista, archived at Wayback Machine, citing a capture dated February 20, 1999, https://web.archive.org/web/19990220030615/http://www.arribavista.com/supportfiles/copyright.asp.

19. "Disclaimer," Arriba Vista, archived at Wayback Machine, citing a capture dated February 20, 1999, https://web.archive.org/web/19990220044714/http://www.arribavista.com/supportfiles/disclaimer.asp.

20. "Copyright/Terms of Use."

21. "About Steven Krongold," Krongold Law, archived at Wayback Machine, citing capture dated February 28, 2024, https://web.archive.org/web/20240228033956/https://krongoldlaw.com/steve-krongold/.

22. Complaint, 5–6.

23. Complaint, 6.

24. "Judy Jennison," Perkins Coie, https://www.perkinscoie.com/en/professionals/judy-jennison.html.

25. Brief of Defendant-Appellee Ditto.Com, 11, copy in author's possession.

26. Kelly v. Arriba Soft Corp., 77 F.Supp. 2d 1116, 1119 (C.D. Cal. 1999).

27. Kelly v. Arriba (1999), 1121.

28. Kelly v. Arriba (1999), 1119.

29. Leslie A. Kelly, quoted in "Kelly Wins in Kelly v. Arriba Soft," NetCopyrightLaw.com, archived at Wayback Machine, citing a capture dated October 13, 2002, https://web.archive.org/web/20021013201743/http://netcopyrightlaw.com/kellyvarribasoft.asp.

30. "The Code of Fair Practice for the Graphic Communication Industry," Graphic Artists Guild, archived at Wayback Machine, citing a capture dated October 29, 2000, https://web.archive.org/web/20001029020445/http://www.gag.org/about/fair_code.html.

31. "Ask First Campaign," Graphic Artists Guild, archived at Wayback Machine, citing a capture dated November 9, 2000, https://web.archive.org/web/20001109231500/http://www.gag.org/about/ask_first.html.

32. Dana Blankenhorn, "The Dumbest Lawsuit in Web History," Clickz, February 23, 2000, https://www.clickz.com/the-dumbest-lawsuit-in-web-history/74179/ (URL no longer active).

33. Julia Ptasznik, "This Just Pisses Me Off," Visual Arts Trends, archived at Wayback Machine, citing a capture dated October 26, 2002, https://web.archive.org/web/20021026145601/http://www.visualartstrends.com/Ea/Ea4/eS10-pissesmeoff.html.

34. Katrina Trinich, quoted in Ptasznik, "This Just Pisses Me Off."

35. " 'This Just Pisses Me Off' Receives a 'Ditto' from Our Readers," Visual Arts Trends, archived at Wayback Machine, citing a capture dated June 13, 2002, https://web.archive.org/web/20020613172444/http://www.visualartstrends.com/Ea/Ea5/eS11-ditto.html; "Who We Are in Brief: Welcome to VisualArtsTrends.com," Visual Arts Trends, archived at Wayback Machine, citing a capture dated June 8, 2002, https://web.archive.org/web/20020608203529/http://www.visualartstrends.com/iB/iB.html.

36. Brief for Plaintiff-Appellant, July 19, 2000.

37. "Charles D. Ossola: Special Counsel," Hunton Andrews Kurth, https://www.huntonak.com/en/people/charles-ossola.html.

38. LA News Serv. v. KCAL-TV Channel 9, 108 F.3d 1999 (9th Cir. 1997).

39. Reply Brief for Leslie A. Kelly, October 4, 2000.

40. Kelly v. Arriba Soft Corp., 280 F.3d 934 (9th Cir. 2002).

41. Kelly v. Arriba (2002), 941.

42. Kelly v. Arriba (2002), 941.

43. "EFF Defends Internet Linking: Asks Court to Rehear Ditto.com Case," Electronic Frontier Foundation, February 27, 2002, archived at Wayback Machine, citing a capture dated August 2, 2002, https://web.archive.org/web/20020802023843/http://www.eff.org/IP/Linking/Kelly_v_Arriba_Soft/20020227_eff_pr.html.

44. Brief of the Electronic Frontier Foundation as Amicus Curiae in Support of Petition for Panel Rehearing and Rehearing En Banc by Defendant-Appellee Ditto.Com, Inc. (Formerly Arriba Soft Corporation), Kelly v. Arriba, February 27,

2002, reposted by Electronic Frontier Foundation, archived at Wayback Machine, citing a capture dated August 2, 2002, https://web.archive.org/web/20020802025051/http://www.eff.org/IP/Linking/Kelly_v_Arriba_Soft/20020227_eff_amicus_brief.html#IV.

45. Brief of the American Society of Media Photographers et al., Kelly v. Arriba, March 26, 2002, reposted by NetCopyrightLaw, archived at Wayback Machine, citing a capture dated July 8, 2004, https://web.archive.org/web/20040708153752/http://netcopyrightlaw.com/pdf/ASMPbrief03262002.pdf.

46. Default Judgment, Kelly v. Arriba, March 17, 2004, reposted by NetCopyrightLaw, archived at Wayback Machine, citing a capture dated September 28, 2007, https://web.archive.org/web/20070928222214/http://netcopyrightlaw.com/pdf/kellyvarribasoftjudgemento3182004.pdf.

47. Andy Warhol Found. for the Visual Arts, Inc. v. Goldsmith, 598 U.S. 508 (2023).

CHAPTER EIGHT. BAD POEMS AND FREE PORN

1. Siva Vaidhyanathan, *The Googlization of Everything (And Why We Should Worry)* (Oakland: University of California Press, 2012), 2.

2. Videotape Deposition of Blake A. Field in Declaration of William O'Callaghan in Support of Google's Motion for Summary Judgment, Field v. Google, Inc., No. CV-S-04-0413-RCJ-GWF, Doc. 52, 8 (D. Nev. November 7, 2005).

3. Blake A. Field, "Good Tea," URL to Google cached page, Exhibit 2 in Declaration of O'Callaghan at 37:1–2.

4. Blake A. Field, "Antiperspirant," URL to Google cached page, Exhibit 2 in Declaration of O'Callaghan at 37:25–26.

5. See, e.g., Certificate of Registration, "Good Tea," TXu1-130-900, Exhibit A, First Amended Complaint for Copyright Infringement, Jury Demand, Field v. Google, Inc., No. CV-S-04-0413-RCJ-LRL, Doc. 5-1568, 8 (D. Nev. May 25, 2004).

6. Declaration of Blake A. Field in Declaration of O'Callaghan at 8:19.

7. Complaint for Copyright Infringement, Jury Demand, Field v. Google, Inc., No. CV-S-04-0413-RCJ-LRL (D. Nev. April 6, 2004).

8. Blake A. Field, "Good Tea," 2004, Exhibit B in First Amended Complaint at 13-1.

9. Complaint for Copyright Infringement, Field v. Google, Doc. 1, 4:22–23 (D. Nev. April 6, 2004).

10. First Amended Complaint for Copyright Infringement at 5:17.

11. Michael Kwun to Blake Field, April 16, 2004, Exhibit 7 in Declaration of O'Callaghan, 25.

12. Memorandum of Points and Authorities in Support of Google Inc.'s Motion for Summary Judgment, Field v. Google, Inc., No. CV-S-04-0413-RCJ-GWF, Doc. 56, 1:9–11 (D. Nev. November 8, 2005).

13. Declaration of Bill Brougher in Support of Google's Motion for Summary Judgment, Field v. Google, Inc., No. CV-S-04-0413-RCJ-GWF, Doc. 56, 1:9–11 (D. Nev. November 15, 2005).

14. Declaration of Brougher, 1.

15. Declaration of Brougher, 1.

16. Declaration of Brougher at 2:1–4.

17. Declaration of Brougher at 2:7–8.

18. Declaration of Brougher at 3:10–11.

19. Martijn Koster, "A Standard for Robot Exclusion," www.robotstxt.org, archived at Wayback Machine, citing a capture dated September 4, 2005, https://web.archive.org/web/20050904213853/http://www.robotstxt.org/wc/norobots.html.

20. "Remove Content from Google's Index," Google, archived at Wayback Machine, citing a capture dated September 13, 2005, https://web.archive.org/web/20050913232556/google.com/remove.html.

21. Declaration of Brougher, 3–4.

22. Expert Report of Dr. John R. Levine, Exhibit 1 in Declaration of Dr. John R. Levine in Support of Google's Motion for Summary Judgment, Field v. Google, Inc., No. CV-S-04-0413-RCJ-LRL, Doc. 54, 12:16–17 (D. Nev. November 8, 2005).

23. Expert Report of Levine, 7:9.

24. Expert Report of Levine, 10:19–23.

25. Expert Report of Levine, 11:1–2.

26. Field v. Google, Inc., 412 F.Supp.2d 1106, 1115 (D. Nev. 2006).

27. Religious Tech. Center v. Netcom On-Line Comm., 907 F. Supp. 1361, 1369 (N.D. Cal. 1995).

28. CoStar Group, Inc. v. LoopNet, Inc., 373 F.3d 544, 555 (4th Cir. 2004).

29. Field v. Google, 1114.

30. Campbell v. Acuff-Rose, cited in Field v. Google, 1118.

31. Field v. Google, 1118.

32. @searchliaison, "Hey, catching up," posted February 1, 2024, https://x.com/searchliaison/status/1753156161509916873.

33. Steven Pinker, quoted in Alvin Powell, "Will ChatGPT Supplant Us as Writers, Thinkers?," *Harvard Gazette*, February 14, 2023, https://news.harvard.edu/gazette/story/2023/02/will-chatgpt-replace-human-writers-pinker-weighs-in/.

34. Norman Zadeh (norman-zadeh-b48b0a176), LinkedIn profile, https://www.linkedin.com/in/norman-zadeh-b48b0a176; Alicia McElhaney, "The Poker-Playing Adult-Magazine Mogul behind One of the Biggest Trading Competitions Is Back," *Institutional Investor*, October 29, 2020, https://www.institutionalinvestor.com/article/b1potnysm2zlzt/The-Poker-Playing-Adult-Magazine-Mogul-Behind-One-of-the-Biggest-Trading-Competitions-Is-Back.

35. Jonathan G. Katz, "Order Instituting Public Administrative and Cease-and-Desist Proceeding, Making Findings, Imposing Remedial Sanctions, and Ordering Respondents to Cease and Desist," July 31, 1998, https://www.sec.gov/litigation/admin/337560.txt.

36. McElhaney, "Poker-Playing Adult-Magazine Mogul."

37. Norman Zada, quoted in Dana Calvo, "Naturally, He Will Try His Breast: A Nudie Mag sans Implants," Associated Press, reposted by Rohit Khare in "Perfect 10," message to Friends of Rohit Khare mailing list, February 12, 1998, https://www.xent.com/FoRK-archive/feb98/0168.html.

38. Zada, quoted in Calvo, "Naturally, He Will Try His Breast."

39. "All Natural au Naturel," *US News*, reposted by Khare in "Perfect 10."

40. Perfect 10, Inc. v. Google, Inc., 416 F. Supp. 2d 828, 832 (C.D. Cal. 2006).

41. Order Granting Defendants' Motion for Attorneys' Fees and Costs in Part and Awarding Defendants $5,213,117.06 in Attorneys' Fees and $424,235.47 in Non-Taxable Costs, Docket 644, Perfect 10, Inc. v. Giganews, Inc., No. CV 11-07098-AB (SHx), Doc. 684, 17:22–24 (C.D. Cal. March 24, 2015), http://cdn.arstechnica.net/wp-content/uploads/2015/03/P10.attys_.fees_.order_.pdf.

42. Order Granting Defendants' Motion at 17–18, 18:1–3.

43. Order Granting Defendants' Motion at 18:1–3, 18:20–21.

44. Order Granting Defendants' Motion at 18:20–21.

45. Order on Motion for Partial Summary Judgment, Perfect 10, Inc. v. Yandex N.V., No. C 12-01521 WHA (N.D. Cal. May 7, 2013), https://scholar.google.com/scholar_case?case=6911966526022823; Order Granting Partial Summary Judgment, Perfect 10, Inc. v. Yandex N.V., No. C 12-01521 WHA (N.D. Cal. July 12, 2013), reposted by Santa Clara Law Digital Commons, https://digitalcommons.law.scu.edu/historical/456.

46. Order on Motion for Partial Summary Judgment at 5.

47. First Brief on Cross Appeal, Perfect 10, Inc. v. Google, Inc., No. 10-56316, Docket 25-1, 7 (9th Cir., December 14, 2010), 6–7.

48. First Brief on Cross Appeal, 6.

49. Response Brief of Defendant Appellee, Perfect 10, Inc. v. Google, Inc., No. 10-56316, Docket 25-1, 7 (9th Cir. December 14, 2010).

50. Response Brief of Defendant Appellee, 8.

51. Image in the public domain, *President Gerald R. Ford Watching Pelé Head a Soccer Ball in the Rose Garden*, US National Archives, June 28, 1975, image hosted on National Archives and Records Administration Domain Archive, https://nara.getarchive.net/media/president-gerald-r-ford-watching-pele-head-a-soccer-ball-in-the-rose-garden-17b221.

52. Perfect 10 v. Google, 839.

53. Perfect 10 v. Google, 839, emphasis in original.

54. Perfect 10 v. Google, 839, emphasis in original.

55. Perfect 10 v. Google, 839, emphasis in original.

56. Perfect 10 v. Google, 840.

57. Hunley v. Instagram, LLC, D.C. No. 3:21-cv-03778-CRB (9th Cir. July 17, 2023), https://cdn.ca9.uscourts.gov/datastore/opinions/2023/07/17/22-15293.pdf.

58. Meg Leta Jones, "Consent Code and Default Drama," in *Just Code: Power, Inequality, and the Global Political Economy of IT* (Baltimore: Johns Hopkins University Press, forthcoming).

59. "Perfect 10 v. Google," Electric Frontier Foundation, https://www.eff.org/cases/perfect-10-v-google.

CHAPTER NINE. THE BIGGEST LIBRARY

1. Helen Nissenbaum, "A Contextual Approach to Privacy Online," *Daedelus* 140, no. 4 (2011): 32–48, available at https://direct.mit.edu/daed/article/140/4/32/26914/A-Contextual-Approach-to-Privacy-Online.

2. Brewster Kahle, "Archiving the Internet," November 4, 1996, submitted to *Scientific American* for March 1997 issue, archived at Wayback Machine, citing a capture dated December 30, 2006, https://web.archive.org/web/20061230164112/https://www.uibk.ac.at/voeb/texte/kahle.html.

3. Sayak Boral, "How to Bypass Paywalls of Leading News Websites," MakeTechEasier, updated October 5, 2022, archived at Wayback Machine, citing a capture dated February 1, 2023, https://web.archive.org/web/20230201231801/https://www.maketecheasier.com/bypass-paywalls-of-leading-news-websites/; magnolia1234, Bypass Paywalls Clean for Chrome, GitHub repository owned by qnoum, archived at Wayback Machine, citing a capture dated February 13, 2023, https://web.archive.org/web/20230213101039/https://github.com/qnoum/bypass-paywalls-chrome-clean-magnolia1234.

4. "$5 Million Gift Supports Harvard's Open Collections Program: Enables Harvard Libraries to Make Collections Available Worldwide," Harvard University Library, July 12, 2004, archived at Wayback Machine, citing a capture dated August 3, 2004, https://web.archive.org/web/20040803001928/http://hul.harvard.edu/publications/ocfunding040712.html.

5. "Harvard Libraries and Google Announce Pilot Digitization Project with Potential Benefits to Scholars Worldwide," Harvard University Library, December 13, 2004, archived at Wayback Machine, citing a capture dated December 31, 2004, https://web.archive.org/web/20041231055824/http://hul.harvard.edu/publications/041213news.html.

6. "About Google Print (Beta)," Google, archived at Wayback Machine, citing a capture dated October 21, 2004, https://web.archive.org/web/20041021010553/http://print.google.com/googleprint/about.html.

7. "Partner Program—an Online Book Marketing Program," Google Book Search (Beta), archived at Wayback Machine, citing a capture dated December 1, 2005, https://web.archive.org/web/20051201012451/http://books.google.com/googlebooks/publisher.html.

8. "Program for Publishers," Google Print (Beta), archived at Wayback Machine, citing a capture dated September 17, 2005, https://web.archive.org/web/20050917025606/https://print.google.com/publisher/?hl=en_US.

9. Patricia Schroeder, quoted in "Google Library Project Raises Serious Questions for Publishers and Authors," Press Release, Association of American Publishers, August 12, 2005, quoted in Jonathan Band, "The Google Print Library Project: A Copyright Analysis," Policy Bandwidth, 2005, archived at Wayback Machine, citing a capture dated September 23, 2005, https://web.archive.org/web/20050923094408/https://www.policybandwidth.com/doc/googleprint.pdf.

10. Class Action Complaint: Jury Trial Demanded, The Author's [*sic*] Guild v. Google Inc., No. 05 CV 8136 (S.D.N.Y. September 25, 2005), reposted by Electronic Frontier Foundation, archived at Wayback Machine, citing a capture dated October 13, 2005, https://web.archive.org/web/20051013072341/http://eff.org/legal/cases/authorsguild_v_google/complaint.pdf.

11. "Google Checks Out Library Books," Google, archived at Wayback Machine, citing a capture dated December 14, 2004, https://web.archive.org/web/20041214233227/http://print.google.com/googleprint/library.html.

12. "Library Project," Google Print (Beta), archived at Wayback Machine, citing a capture dated September 24, 2005, https://web.archive.org/web/20050924011456/http://print.google.com/googleprint/library.html.

13. Kelly v. Arriba Soft Corp., 336 F.3d 811, 818 (9th Cir. 2003).

14. Kelly v. Arriba, 818.

15. Band, "Google Print Library Project."

16. "Can We Trust Google with Our Secrets?," cover, Time, February 20, 2006, posted on Time archive, archived at Wayback Machine, citing a capture dated February 21, 2006, https://web.archive.org/web/20060221041948/http://www.time.com/time/covers/0%2C16641%2C1101060220%2C00.html.

17. "Can We Trust Google with Our Secrets?"

18. Adi Ignatius, "In Search of the Real Google," Time, February 20, 2006, https://content.time.com/time/subscriber/article/0,33009,1158961,00.html.

19. Ignatius, "In Search of the Real Google."

20. Vint Cerf, quoted in Ignatius, "In Search of the Real Google."

21. Google Books Search, Blogspot, http://booksearch.blogspot.com/search?updated-max=2006-08-17T10:00:00-07:00&max-results=10.

22. Laura Moody, quoted in Pam Saenger, "Pick Up Your Library Pass to Google," Blogspot post to Google Books Search blog, May 30, 2006, http://booksearch.blogspot.com/2006/05/pick-up-your-library-pass-to-google.html.

23. Saenger, "Pick Up Your Library Pass to Google."

24. John Wilkin, "Welcome to the University of California Libraries," Google Official Blog, August 9, 2006, https://googleblog.blogspot.com/2006/08/welcome-to-university-of-california.html.

25. Jennifer Colvin, "UC Libraries Partner with Google to Digitize Books: UC Becomes the Newest Partner in the Google Books Library Project," University of California Office of the President, August 9, 2006, archived at Wayback Machine, citing a capture dated September 1, 2006, https://web.archive.org/web/20060901082958/http://www.universityofcalifornia.edu/news/2006/aug09.html.

26. Susan Crawford, "Why Google Is Right," blog post to Susan Crawford blog, September 21, 2005, archived at Wayback Machine, citing a capture dated October 25, 2006, https://web.archive.org/web/20061025152815/http://scrawford.blogware.com/blog/_archives/2005/9/21/1248170.html.

27. Lawrence Lessig, "Google Sued," Lessig blog, September 22, 2005, archived at Wayback Machine, citing a capture dated October 20, 2006, https://web.archive.org/web/20061020032955/http://www.lessig.org/blog/archives/003140.shtml.

28. Arielle Reinstein, "A Look Back on 2006," Inside Google Book Search, January 4, 2007, archived at Wayback Machine, citing a capture dated July 1, 2009, https://web.archive.org/web/20090701095455/http://booksearch.blogspot.com/2007/01/look-back-on-2006.html; Jodi Healy, "University of Texas at Austin Becomes Our Latest Library Partner," Inside Google Book Search,

January 19, 2007, archived at Wayback Machine, citing a capture dated March 31, 2007, https://web.archive.org/web/20070331232238/http://booksearch.blogspot.com/2007/01/university-of-texas-at-austin-becomes.html.

29. Cindy Cohn, quoted in "Google Book Search: News and Views—Legal Analysis," Google Book Search (Beta), archived at Wayback Machine, citing a capture dated November 23, 2006, https://web.archive.org/web/20061123213041/http://books.google.com/googlebooks/newsviews/legal.html.

30. Fred von Lohmann, "Google Book Search Settlement: A Reader's Guide," Electronic Frontier Foundation, October 31, 2008, updated August 2009, https://www.eff.org/deeplinks/2008/10/google-books-settlement-readers-guide.

31. Attachment I of Settlement Agreement, Exhibit 1 of Declaration of Michael J. Boni in Support of Plaintiffs' Motion for Preliminary Settlement Approval, Authors Guild, Inc. v. Google, Inc., No. 05 CV 8136-JES, Doc. 56, 248–249 (S.D.N.Y. October 28, 2008). Copy in the author's possession.

32. Statement of Interest of the United States of America Regarding Proposed Amended Settlement Agreement, Authors Guild, Inc. v. Google, Inc., No. 05 Civ. 8136(DC), Doc. 922 (S.D.N.Y. February 4, 2010), US Department of Justice, Antitrust Division, updated October 18, 2023, https://www.justice.gov/atr/case-document/statement-interest-united-states-america-regarding-proposed-amended-settlement.

33. Statement of Interest of the United States of America Regarding Proposed Amended Settlement Agreement.

34. James Grimmelmann, quoted in Diane Bartz, "U.S. Justice Department Looks into Google Books Deal," Reuters, April 28, 2009, https://www.reuters.com/article/us-google-books-antitrust/u-s-justice-department-looks-into-google-books-deal-idUSTRE53R8DO20090429.

35. "Yahoo! Inc. and Google Inc. Abandon Their Advertising Agreement: Resolves Justice Department's Antitrust Concerns, Competition Is Preserved in Markets for Internet Search Advertising," Press Release, US Department of Justice, November 5, 2008, https://www.justice.gov/archive/opa/pr/2008/November/08-at-981.html.

36. "Justice Department Sues Google for Monopolizing Digital Advertising Technologies," Press Release, US Department of Justice, Office of Public Affairs, January 24, 2023, https://www.justice.gov/opa/pr/justice-department-sues-google-monopolizing-digital-advertising-technologies.

37. Motoko Rich, "Google Hopes to Open a Trove of Little-Seen Books," *New York Times,* January 4, 2009.

38. Rich, "Google Hopes to Open a Trove of Little-Seen Books."

39. "Google Books Privacy Policy," Google Books (Beta), September 3, 2009, archived at Wayback Machine, citing a capture dated September 18, 2009, https://web.archive.org/web/20090918155813/http://books.google.com/googlebooks/privacy.html.

40. Maureen Dowd, "Dinosaur at the Gate," *New York Times,* April 14, 2009.

41. Dowd, "Dinosaur at the Gate."

42. Michael Barbaro and Tom Zeller Jr., "A Face Is Exposed for AOL Searcher No. 4417749," *New York Times*, August 9, 2006.

43. United States v. Rumley, 345 U.S. 41, 57 (1953), cited in Privacy Authors and Publishers' Objection to the Proposed Settlement, Authors Guild v. Google, No. 05 CV 8136-DC, Doc. 325, 12 (S.D.N.Y. September 9, 2009); Barbaro and Zeller Jr., "Face Is Exposed."

44. Denver Area Ed. Telecommunications Consortium, Inc. v. Federal Commc'ns Comm'n, 518 U.S. 727, 754 (1996), cited in Privacy Authors and Publishers' Objection to the Proposed Settlement, 12.

45. In re Grand Jury Subpoena to Kramerbooks and Afterwords, Inc., 26 Media L. Rep. (BNA) 1599, 1600 (D.D.C. 1998), cited in Privacy Authors and Publishers' Objection to the Proposed Settlement, 13.

46. Jonathan Lethem, quoted in "Re: Privacy Concerns about to [*sic*] the Google Book Settlement," letter and e-mail to Daralyn J. Durie and Joseph C. Gratz of Durie Tangri Lemley Roberts & Kent LLP, Electronic Frontier Foundation, October 6, 2009, https://www.eff.org/document/group-letter-privacy.

47. Opinion, Authors Guild, Inc. v. Google, Inc., No. 05 Civ. 8136(DC), Doc. 971 (S.D.N.Y. March 22, 2011), 21, emphasis in original, archived at Wayback Machine, citing a capture dated March 26, 2011, https://web.archive.org/web/20110326025439/http://thepublicindex.org/docs/amended_settlement/opinion.pdf.

48. Authors Guild, Inc. v. Google, Inc., 954 F.Supp. 2d 282, 291 (S.D.N.Y. 2013).

49. Authors Guild v. Google (2013), 291.

50. Complaint, N.Y. Times Co. v. Microsoft Corp., 1:23-cv-11195 (S.D.N.Y., Dec. 27, 2023).

EPILOGUE

1. "Digital Copyright Act of 2021: 12/18 Discussion Draft for Stakeholder Comments Only (Not Final)," December 18, 2020, https://www.tillis.senate.gov/services/files/97A73ED6-EBDF-4206-ADEB-6A745015C14B.

2. Google v. Oracle America, 141 S. Ct. 1183 (2001), 1188.

3. Andy Warhol Found. for the Visual Arts, Inc. v. Goldsmith, 598 U.S. __ (2023).

ACKNOWLEDGMENTS

This book was made possible by the generous financial support of several organizations. The Sloan Foundation made possible most of the research that went into this project through a grant from its Public Understanding of Science Program. The Hoover Institution at Stanford University, where I held a National Fellowship, provided me with unrestricted writing time for a year. I also received research support from the Smithsonian National Museum of American History, the National Endowment for the Humanities, and the Center for Intellectual Property and Innovation Policy.

I have been very lucky to have a wonderful scholarly community that has supported my work ever since I was a graduate student. I continue to be in awe of how everything I learned from my PhD mentor, Daniel Kevles, shapes every aspect of my work. This book is very different from the kinds of history he taught me to write, but I hear his voice in everything I do as a professor and will always hope to make him proud. I remember Bettyann Kevles very dearly. Naomi Lamoreux and Mario Biagioli have had a lasting influence on how I work and think and have shown me how to be kind and tough at the same time. I am also very grateful to Gerard Alberts, Bill Aspray, Michael Barany, Martin Campbell-Kelly, Steph Dick, Nathan Ensmenger, Richard Haber, Tom Haigh, David Hemmendinger, Eric Hintz, Richard John, Matthew Jones, Meg Leta Jones, Eden Medina, Mary Mitchell, Adam Mossoff, Kelly O'Donnell, Liz Petrick, Andy Russell, and Rebecca Slayton.

Several anonymous reviewers have provided me with priceless feedback and support along the way.

Jeff Yost has been a dear friend and a close collaborator for many years. His kindness, openness, and generosity have touched the lives and work of countless scholars in our field. Our many projects together have shaped every aspect of my engagement with the social studies of information technology. Our conversations are incredibly generative, as we have the same overall scholarly goals and priorities, but we have wildly different and yet mutually complementary analytical instincts. I am honored to have been appointed a Research Fellow of the Charles Babbage Institute alongside Rebecca Slayton and a few other prominent scholars, and to call Jeff a dear friend.

My approach to copyright has developed slowly since my graduate school years, but in 2020 the Classical Liberal Institute at NYU organized a summerlong symposium that helped me start putting things together in my own way. Special thanks to Richard Epstein, Adam Mossoff, Mario Rizzo, and Bowman Heiden for having me. I am also grateful to the other participants of the workshop, especially Chris Beauchamp, Sean Bottomley, Oren Bracha, Kara Swanson, and John Duffy. What I learned from meeting with them helped me expand and revise approaches I had started to develop when I was writing about patents and meeting with folks at the Hoover IP2 conferences organized by Naomi Lamoreaux and Steve Haber. I am grateful to the 2020–2021 cohort of National Fellows at the Hoover Institution for many enlightening conversations during our year together. At Yale University Press, Bill Frucht has been a blessing of an editor, Mary Pasti coordinated the process, and Joe Calamia (now at Chicago) believed in this project when there was little more to show for it than a proposal. Laura Jones Dooley helped me prepare the manuscript for publication.

I have been immensely lucky to belong to the Science and Technology Studies Department at UC Davis. My colleagues are a delight, and they have pushed me in intellectual directions that I couldn't have imagined when I joined them eight years ago. The intellectual flexibility and openness they have instilled in me have become two of my core values as a scholar. It is also wonderful to have had the opportunity to develop close friendships along the way. Thank you, Finn Brunton, Tim Choy, Marisol de la Cadena, Joe Dumit, Tim Lenoir, Emily Merchant, and Colin Milburn. And welcome to our two newest colleagues, Kayleigh Perkov and Natalia Duong!

At UC Davis, I have benefitted from the work of several undergraduate research interns. Even if they didn't work on copyright, they gave me the space I needed to think about law and technology. Manda McElrath, Emily Barneond, Marshall Comia, Andreas Godderis, Chris Andoh, Kacey Huang, Bohan Xiao, and Stephen Fujimoto all contributed to my thinking. They took deep dives into the Wayback Machine and court records, helped me create archives of news and blog posts, helped me code hundreds of pages of legal documents, and showed me what kinds of stories about law matter in their own lives. I am also grateful to students in my "Digital Law" and "Internet Copyright" classes, whose class participation and papers helped me figure out what's most exciting about these topics. My star mentee and intern, Nico Wen-Li Ong, played a major role in helping me finish this book by helping me format the citations.

On a more personal note: Folks in academia don't talk about mental health very often, but I wrote this book during the worst mental health crisis I have ever had. From 2021 to 2023, I experienced a paralyzing depression. At one point I couldn't even open my eyes or recall information. I was able to recover thanks to a combination of exercise, therapy, medication, self-care, love, and a systematic reevaluation of my life. For this, I'd like to thank my friends and family, especially Vincent, my mother, and Jeff Yost. Our dogs, Kupo and Luna, also did their part. Everyone in my department treated me with kindness and love, helping me find joy in coming into the office to think, write, and teach. I am also grateful to my therapist, Mark Corey, who helped me find my way out and is still helping me reflect on what it means for me to live a good life. If you need help, I hope you will speak with someone—anyone.

My friends and family have been with me from the very beginning of this book. My husband, Vincent Cheng, cheered me on during those days when all I did was go into a coffee shop and write until it was nighttime. My mother, Dora—who in her seventies is acing her way through law school— checked in on me regularly and gave me the privilege of developing an intellectual relationship with her. My dear father continues to watch over us and guide me from beyond. Marisol de la Cadena and Steve Boucher have welcomed me and Vincent into their lives, and I could not imagine life in California without them. Joe Dumit always helps me remember that scholars can be playful, and he is by far the best PokémonGo mentor in the world. Cody and Jessica Winters helped us feel at home in Sacramento. At our local coffee shop, Sammy helped me feel welcome and energized. Sarit Monge, Lily Zeng, and Marijo de la Cadena have known me for a very long time and are my sisters. My brother, Gilberto (the only person I've ever called "Bro"), and his wife, Rosy, have, without fail, helped me become a kinder person. And my niece, Celeste, will always be the coolest person I know.

Thank you.